ANCIENT DNA

ELIZABETH D. JONES

Ancient DNA

THE MAKING OF A

CELEBRITY SCIENCE

Yale UNIVERSITY PRESS   NEW HAVEN AND LONDON

Published with assistance from the Mary Cady Tew Memorial Fund.

Yale University Press books may be purchased in quantity for educational, business, or promotional use. For information, please e-mail sales.press@yale.edu (U.S. office) or sales@yaleup.co.uk (U.K. office).

Set in Scala type by IDS Infotech, Ltd.
Printed in the United States of America.

ISBN 978-0-300-24012-2 (hardcover : alk. paper)
Library of Congress Control Number: 2021939551
A catalogue record for this book is available from the British Library.

This paper meets the requirements of ANSI/NISO Z39.48-1992 (Permanence of Paper).

10 9 8 7 6 5 4 3 2 1

For
Patrick,
Dawson,
and Colton

CONTENTS

My interest in paleontology, and in particular the search for DNA from ancient and extinct organisms, began over a decade ago as an undergraduate majoring in history and philosophy at North Carolina State University. As part of my studies, I was required to take a science elective. I chose the course "Dinosaurian World" because, like many other students, I thought it would be a walk in the park. Dr. Mary Schweitzer taught the course, which was anything but easy. I loved it, though. I spent the next several years digging up dinosaurs in the badlands of Montana, prepping fossils in the lab, and taking a number of other classes on dinosaur anatomy, physiology, and evolution.

After class one day, I asked Dr. Schweitzer a question. I had seen on television that one of the major science channels was airing an upcoming episode on a scientist who had discovered evidence of ancient proteins from a dinosaur. I was planning on watching it but I wanted to know if she knew of this work, and if so, what she thought of it. Ancient dinosaur proteins? Really? She smiled and said to watch it first and that we could talk about it later. So, I did and was mortified when I realized that Dr. Schweitzer was the star of the show, the scientist suggesting she had extracted evidence of ancient proteins from a 60-million-year-old *Tyrannosaurus rex*.

Dr. Schweitzer's findings, published across top-tier academic journals from *Science* to *Nature*, were nothing short of extraordinary. As such, they were quickly and frequently followed by media attention and public speculation. As I learned more about this work, I learned that she was criti-cized just as much, if not more, than she was celebrated for these pioneering efforts in molecular paleontology. Since the early 1990s, she had been inves-tigating the limits of molecular preservation in fossils as a student studying with Jack Horner, a maverick dinosaur paleontologist and scientific consul-tant on Steven Spielberg's cinematic production of Michael Crichton's *Jurassic Park*. Over the years, she had acquired loyal supporters as well as vehement skeptics who resisted the idea that molecules, whether proteins or DNA, could survive intact for millions of years.

My exposure to this paleontological controversy, and the very public nature of it, is what inspired this book. Consequently, the book is a historical examination of the search for molecules from fossils, specifically the search for DNA from fossils, as it progressed from an idea of science fiction into a research reality from the early 1980s to today. In these pages, I explore the controversy and celebrity that not only followed it but actually came to direct and define, in quite remarkable ways, the formation of a new scientific field, now widely known by both the scientific community and broader public as the field of "ancient DNA research."

ACKNOWLEDGMENTS

There are many who have inspired and mentored me. First, I thank my paleontology professors, Mary Schweitzer at North Carolina State University and Gregory Erickson at Florida State University, who encouraged me to pursue paleontology and its intersections with the science of genetics. I also acknowledge Will Kimler, North Carolina State University, and Paul Brinkman, North Carolina Museum of Natural Sciences, who introduced me to the history of science, and Sam Schmidt, who keeps me updated on the latest fossil discoveries. I am especially grateful to the History and Philosophy of Science Program at Florida State University and Fritz Davis, who advised my master's research, which became the foundation for this book.

Next, the extensiveness of this research was made possible with funding from University College London (UCL) through three separate scholarships: a UCL Graduate Research Scholarship, a UCL Overseas Research Scholarship, and a UCL Cross-Disciplinary Training Scholarship. I am also grateful to UCL's Department of Science and Technology Studies, the British Society for the History of Science, the History of Science Society, and the Division of Paleontology at the American Museum of Natural History for funding toward research travel to conduct interviews and deliver presentations.

I also want to acknowledge the paleontologists and geneticists in UCL'S Department of Genetics, Evolution and Environment who have greatly

informed and critiqued this research as it was developed. A special thanks
goes to Marcela Randau Carvalho Burgess, Andrew Cuff, Thomas Halliday,
Ryan Felice, Aki Watanabe, and Carla Bardua of Anjali Goswami's Lab.
Thanks also go to Mark Thomas and the Molecular and Cultural Evolution
Lab, especially Zuzana Faltyskova, Anna Rudzinski, Elizabeth Gallagher,
Katherine Brown, Catherine Walker, Pascale Gerbault, Lucy van Dorp, Yoan
Diekmann, Adrian Timpson, Stuart Peters, and David Diez del Molino. I am
further appreciative for the opportunity to work with Selina Brace, Ian
Barnes, and Tom Booth from the Natural History Museum as part of the
UCL Cross-Disciplinary Scholarship Training.

From UCL's Department of Science and Technology Studies, I am
thankful for the support from staff, faculty, and colleagues. Specifically,
Raquel Velho, Erman Sozudogru, Oliver Marsh, Toby Friend, and Julia
Sanchez-Dorado have offered valuable insights and encouragement. Further,
I want to acknowledge Joe Cain, my doctoral supervisor, whose intellectual
guidance was central to the development of the argument presented here.

Overall, the strength of this book rests in the fact that it was produced in
partnership with scientists, especially those who helped found and further
the field of ancient DNA research. I am indebted to each interviewee for
giving so much of his or her time to the telling and writing of this history. A
number have shared or donated documents from their research, helping me
to create an archive for the future. These interviewees and their memories
are priceless. I hope they see their experiences in these pages.

This book was acquired for Yale University Press by Jean Thomson
Black, and I am beyond grateful for her vision and patience, and the
opportunity to publish with a great team. Michael Deneen, Elizabeth
Sylvia, Philip King, and Margaret Hogan have expertly guided this manuscript through the editorial process. Further, this manuscript has been
much improved thanks to comments from David Sepkoski, and several
anonymous reviewers.

Finally, my career as a historian of science has been possible because of
my family, both the Dobsons and the Joneses, and in particular my parents,
Allen and Martha, who have supported this research through its many
stages. Most of all, I want to acknowledge my best friend and husband,
Patrick. The path to achieving our accomplishments is never a linear or solitary journey, and I have him to thank for helping to bring this book into the
world.

ANCIENT DNA

Introduction

IN 2015, SCIENTISTS FROM UNIVERSITY COLLEGE LONDON and Imperial College London reported the recovery of ancient tissue structures from 75-million-year-old dinosaur bones. *Nature Communications* published the findings on June 9, 2015, the day that *Jurassic World* premiered in Hollywood.[1] The film was released three days later to audiences across sixty countries, generating $550 million worldwide and ranking as the fourth-highest-grossing film in history. The coincidental timing did not go unnoticed. A media article from *The Independent* in London covered the story: "Just days before Steven Spielberg releases his latest dinosaur blockbuster in the shape of *Jurassic World*—the fourth film in the franchise—scientists have announced that they have found evidence of red blood cells and protein preserved in 75-million-year-old dinosaur fossils."[2] There was no evidence for dinosaur DNA, the main motif driving Michael Crichton's and Steven Spielberg's original filming of *Jurassic Park,* but scientists saw evidence of protein preservation. They also saw evidence for what looked like the remnants of blood cells.

This study was not the first to report the potential preservation of biomolecules or soft tissue structures from the days of the dinosaurs. There were earlier reports from the 1990s that offered evidence for this phenomenon too. Further, this 2015 study was not the first to attract media attention or to be framed with the fame of a major blockbuster movie. Indeed, this echoed a particular event reminiscent of the 1990s when Raúl J. Cano and a team of researchers in California announced they had extracted and sequenced DNA from a 120–135-million-year-old insect preserved in amber. Their discovery was reported in *Nature* on June 10, 1993, the day after the first *Jurassic Park* film premiered and the day before its public release worldwide.[3] A *New York Times* article, "DNA from the Age of the Dinosaurs Is Found," acknowledged the timeliness of the publication with the premiere, even though the authors argued it was coincidental.[4] Coincidence or not, over the past three decades, press and public interest have followed the search for molecules in fossils, especially the search for DNA from fossils, extremely closely. Publicity has followed the search for DNA from fossils so closely, in fact, that a whole scientific field has been shaped by its ever growing celebrity and the controversies that ensued from it.

This book documents the development of the relatively young scientific field known as ancient DNA research, revealing the intimate interaction between science and the media in its evolution from a speculative, esoteric idea into a credible, revolutionary approach to the study of evolutionary history. Today, ancient DNA research is a fast-paced, high-cost, high-stakes, and technologically savvy practice that continuously challenges our assumptions about the past and the means through which to discover it. The field has captured professional and popular attention as researchers have searched for multimillion-year-old dinosaur DNA and recovered the genomes of extinct mammoths, ancient diseases, early humans, and our archaic ancestors (the Neanderthals and Denisovans) in an attempt to refine or even rewrite our understanding of how life came to be. Given the profound nature of findings generated by scientists in this field, a history of ancient DNA research offers the opportunity to understand how the study of fossils has impacted evolutionary biology and our conceptions of evolutionary history. Even more interesting, however, is the field's uniquely high-profile character among press and public audiences. A history of ancient DNA research offers the opportunity to understand the intricate interchange between science and media in terms of how the practice, process, and communication of science were in turn influenced by it.

I offer in these pages the first historical, philosophical, and sociological account of the search for DNA from fossils that addresses its disciplinary development over a thirty-year period, from the 1980s to today, in the context of the mass media, celebrity culture, and modern science communication movement. Drawing on historical and archival material, as well as original interviews with more than fifty scientists from North America to Europe and Australia, I explore ancient DNA research's formation and explain its relationship with the media by examining its close connection to de-extinction, the science and technology of resurrecting extinct species. In doing so, I reveal how the search for DNA from fossils developed into a discipline under the influence of intense press and public interest, particularly as this new line of research coincided with and was catapulted into the media spotlight by the book and movie *Jurassic Park*.

As the search for ancient DNA evolved from the 1980s to the present, the researchers involved in it responded to the field's technological challenges and increasing identity as a celebrity science. Over the years, they became concerned about "contamination" in both a literal and figurative sense. Contamination, in the literal and technical understanding of the term, refers to potential exposure of exogenous DNA from the environment, bacteria, and even humans. For example, a specimen's exposure to other DNA sources can happen over time or through handling in a museum collection or lab. This issue of contamination is further heightened by the already degraded and fragmented composition of DNA from ancient and extinct organisms. Therefore, it can be difficult for researchers to determine what genetic sequences, if any, belong to the actual specimen under study. This question of ancient DNA authenticity was a particular problem for scientists in the early days of the discipline's development. Authenticity was a prerequisite for credibility, on which both the success of the practice and the reputation of practitioners promoting it depended.

At the same time, scientists were also worried about the influence of celebrity. Indeed, some researchers, working both inside and outside the discipline, viewed the hype around the search for DNA from fossils as a further and more figurative source of contamination, but one that was no less real. According to some scientists, disproportionate and undeserved media interest or influence had the adverse effect of contaminating their credibility. Interestingly enough, just as media attention seemed to undermine the young field, it empowered it too. Indeed, the media served as a crucial component to its growth in terms of its initial formation and overall identity. Over the years, the press consistently publicized the nascent science, but scientists also purposefully fashioned their own opportunities for attention. This intentional exchange between scientists and the media—specifically around the idea of discovering DNA from some of the world's most ancient and charismatic creatures, such as dinosaurs or mammoths—influenced publication timing, grant funding, research agendas, and professional recruitment. I argue that scientists' concerns about contamination, as related to both ancient DNA authenticity and the influence of celebrity, and the way they articulated those concerns privately as well as publicly, played a fundamental role in driving, even defining, ancient DNA research as a scientific and technological practice in its own right.

ANCIENT DNA RESEARCH

For this book, ancient DNA research is broadly defined as the practice of extracting, sequencing, and analyzing degraded or damaged DNA from long-dead organisms, ranging from hundreds or hundreds of thousands to even millions of years old. Although there is no convincing evidence for the long-term preservation of multimillion-year-old DNA, scientists have searched for it, and some continue to do so, in hopes of eventually accessing deep time. So far, the record for the oldest DNA, and in this case the oldest genome, comes from 1-million-year-old mammoth teeth frozen in the permafrost of northeastern Siberia.[5] Since the specimen reaches all the way back to the million-year mark, this is an especially remarkable achievement, considering that one of the earliest milestone studies in this field involved the recovery of a modest number of DNA sequences from skin and tissue samples of a 140-year-old extinct horse-like mammal that once roamed the plains of South Africa.[6] Reflecting on these studies side by side, it seems strange to refer to such a specimen as ancient or the genetic material as ancient DNA. However, in the writing and telling of the history of ancient DNA research, scientists reference specimens from the recent to the more distant past in their own accounts, namely because one of the primary objectives of this line of work is to better understand the fickle nature of molecular preservation over time.

Indeed, the term "ancient DNA" is not necessarily used to denote the age of the DNA but to highlight characteristic damage patterns that occur as genetic material breaks down after an organism has died. After an organism dies, its body decays due to internal chemical as well as external degradation processes, leaving little more than skeletal remains of a creature long gone. If any DNA is left, it is often chemically altered, resulting in fragmented sequences. This results in a lower quality and quantity of sequences when compared with modern genetic material. Further, the length of time an organism has been left to decay as well as the type of environment in which it has been left to decay play a role in how much DNA may be preserved. Considering this, research into DNA from dead organisms requires specialized skills, knowledge, and technologies to conduct it. Thus, many practitioners are invested in adapting state-of-the-art molecular biological techniques and high-throughput sequencing technologies to optimize the recovery of ancient DNA, be it from an organism a few hundred years old or a few thousand years old.

In addition to the wide temporal range of organisms that fall under the definition of ancient DNA research, there are numerous organisms from which DNA can be extracted. Ancient DNA, for example, can be recovered from plants, animals, humans, and bacteria, and it can be preserved in skin, tissues, and even bone if the bone is not a fully mineralized fossil. An organism's status as a fossil or subfossil matters greatly when considering the possibility that some cellular or molecular structures may be preserved. A fully mineralized fossil is unlikely to preserve DNA, because its organic components have decayed and been replaced by minerals from the surrounding sediment in which the organism died. A subfossil, on the other hand, is only a partially mineralized organism and therefore may retain some of its organic components. If this is the case, it is possible that cellular and molecular components remain intact. On this point, it is also crucial to consider whether the specimen under study exists as a piece of skin, tissue, or bone. These various substances may have different capacities to preserve molecules, which in turn can affect scientists' approach to the isolation and extraction of genetic material.[7] Such differences are important, and I reference specifics regarding what type of specimen is under study and the corresponding findings when it is relevant to illustrating a methodological or conceptual point. Otherwise, I use the broad terms "ancient DNA research," "the search for DNA from ancient and extinct organisms," or "the search for DNA from fossils" interchangeably to refer to the general investigation of degraded and damaged DNA from various sources.

Over the past three to four decades, a number of phrases have been used by scientists, as well as media journalists, to refer to the search for ancient DNA. For example, in the late 1970s and early 1980s, a small group of individuals started talking about the possibility of recovering DNA from long-dead organisms, using a range of terms to do so. They talked about "old

DNA," "extinct DNA," "fossil DNA," and "paleoDNA."[8] They also used the term "ancient DNA," which later became the flagship label when it was adopted and advertised widely by a series of professional conferences on the topic hosted across the United Kingdom in the 1990s.[9] Today, "ancient DNA research" remains the pervasive terminology, but references to "molecular archeology," "molecular paleontology," "archeogenetics," and "paleogenetics" are also common, although they carry their own nuances in usage.[10]

METHODS AND SOURCES

From the 1980s to today, the ancient DNA community of practitioners has come to be international and interdisciplinary in its nature. Although the first labs to promote and popularize this practice originated in the United States and the United Kingdom, there are a number of research centers across North and South America as well as Europe, Asia, and Australia that specialize in this niche line of work. This geographical spread is matched by the disciplinary diversity of the practitioners themselves, coming from various scientific backgrounds, steeped in their own unique set of research questions, methods, and traditions. Consequently, the ancient DNA community is made up of many research subspecialties, and a history of the field can be told from various disciplinary viewpoints. An account of the practice and its practitioners, for example, can be told from a paleontological perspective or with exclusive interest in archeological and anthropological approaches, highlighting the unique contributions, challenges, and controversies in each research focus. In fact, most book-length works on the topic of ancient DNA research have addressed the latter and are firsthand accounts from ancient DNA researchers themselves.[11]

Recently, however, Elsbeth Bösl—a historian of science and the author of *Doing Ancient DNA*—traced the growth of ancient DNA research through a Eurocentric and disciplinary-specific focus on the field's relationship with archeology. Drawing on an extensive review of scientific and popular publications, along with interviews with scientists from German-based institutions, Bösl addressed the nuances of the search for ancient DNA as applied to archeology. Notably, she investigated the controversies geneticists and archeologists encountered as they negotiated the explanatory power and hierarchical nature of their respective data sources, be it genetic sequences or ancient material artifacts, to shed light on cultural, social, and historical hypotheses about human evolution and migration.[12]

In contrast, I have approached the history of ancient DNA research in evolutionary biology much more broadly. In doing so, I have captured the events and experiences of those individuals who identify with the ancient DNA community through a mutual pursuit to discover old molecules from a broad range of sources for the primary purpose of studying the origin, evolution, and migration of extinct and extant organisms across the world. I have also included individuals who entertain the more visionary usages of old molecules, such as bringing extinct creatures back to life. Such a view of the field is not only exciting but necessary considering ancient DNA research's truly interdisciplinary nature and ever increasing popularity with the media. This broad view of the field is essential for more fully under-standing the key ideas, events, individuals, and experiences that have affected the evolution of this field as a whole.

My methods include examination of historical and archival material, as well as original interviews with more than fifty scientists from North America, Europe, and Australia. For the historical material, I draw on scientific publications on ancient DNA research and corresponding media articles from newspapers, magazines, journals, and websites, as well as popular books and films. I have also consulted archival documents from the Allan Wilson Papers in the Bancroft Library at the University of California, Berkeley. Further, I have had access to numerous unpublished documents such as original funding applications, manuscripts, newsletters, correspondences, and memoirs, along with conference and workshop programs, that were provided to me by scientists I interviewed.

As a historian of science, I used oral history methods to inform my interviews with fifty scientists working in and on the periphery of ancient DNA research. The selected interviewees represent researchers from disparate disciplines in evolutionary biology, including paleontology, archeology, anthropology, botany, epidemiology, evolutionary genetics, population genetics, molecular biology, microbiology, and computational biology. These interviewees work in the United States, Canada, England, Ireland, Australia, Germany, Denmark, Sweden, Norway, France, Spain, and Israel. I tried to interview researchers who represent different scientific, epistemic, and generational views regarding the practice of ancient DNA research. Five additional group interviews with doctoral and postdoctoral researchers were also completed. The method for both individual and group interviews was semi-structured in style, on average two hours in length, and resulted in partial transcriptions for analysis. Interviewees were not selected at random but were a sample of the population of researchers in North America, Europe, and Australia.[13]

All scientists consented to interviews under the condition that quotations would be anonymized. Anonymization was chosen in order to protect researchers' identities and reputations, which in turn allowed them to be more candid with sharing their experiences. For the most part, scientists enjoy discussing their achievements as well as the challenges they encounter. They often do so openly. However, science is a social process shaped by personalities, ambitions, political systems, cultural pressures, and capitalistic interests, just as much as it is shaped by our curiosity to understand the world around us. The field of ancient DNA research is a highly competitive one filled with controversy. In this discipline—very much like in the related fields of anthropology, archeology, and paleontology—specimens and money are rare, so competition for access to both is fierce. As such, many ancient DNA researchers felt more comfortable knowing their information would be anonymous, to avoid professional backlash. This is particularly important as most, if not all, interviewees are practicing scientists with their own careers and their students' careers at stake. Therefore, interviewees have been anonymized with a random number. There are times when researchers are identified in association with quotations, but in these cases such quotations come from correspondence with permission to be identified or from previously published articles and documents.

Additionally, I came to this work with the intention of understanding ancient DNA research's development in the context of the mass media, celebrity culture, and the modern science communication movement. Specifically, I was interested in examining the discipline's relationship with the media and its fascination with the idea of resurrecting extinct species. The semi-structured interviews reflect this intention as interviewees were asked about their background and education, professional and theoretical commitments, perspectives on ancient DNA research, perspectives on the media, and views on the science of resurrecting extinct species.

The goal of the interview transcriptions was qualitative and thematic. Throughout the data collection, transcription, and analysis process, I listened for recurring themes in the retelling of the history. I also listened for deviations from shared themes in the form of subtle disagreements to outright controversies. The ancient DNA community is a contentious one, and the researchers working in it openly recognize that the field is full of colorful, ambitious, and provocative personalities. Throughout this work, I have incorporated the points of dispute and attempted to give a voice to every interviewee's experiences. But despite interviewee individuality, there remains a clear, overarching theme about the field's history that was shared among scientists across space and time: the role of contamination as it relates to ancient DNA authenticity and, even more interestingly, as it sometimes relates to celebrity and what some scientists saw as too much or the wrong kind of media attention. All interviewees agreed, though to differing degrees, on the role of the media and scientists' interactions with the media as direct or indirect influences on the development of ancient DNA research.

SCIENCE IN THE SPOTLIGHT

This book is about the interplay between science and the media as the search for DNA from fossils developed from science fiction into a research reality from the 1980s to today. Over the past few decades, ancient DNA research has grown into much more than a public-facing science. I argue it has grown into a celebrity science. This new term—"celebrity science"—is the culmination of my analysis of the historical, archival, and interview evidence detailed in the forthcoming chapters and outlined more fully at the end of the book. Broadly, I define a celebrity science as a subject of science that exists and evolves under intense public interest and extreme media exposure. While media platforms present scientists working in or around the field of interest with opportunities for publicity, scientists are also pragmatic in fashioning their own occasions for public notice. Overall, a celebrity science is the outcome of prolonged publicity; it is the process and product of consistent science-media interactions around a scientific subject over an extended period of time. I suggest that the celebrity science concept helps make sense of ancient DNA research's historical and sociological development into a legitimate scientific practice. I also suggest it can help other scholars make sense of other sciences operating in the media spotlight. Consequently, ancient DNA research is a case study of the making of a celebrity science but is by no means an exclusive example of it.

Four points are important. First, it is necessary to distinguish between publicity and celebrity. The *Oxford English Dictionary* defines "publicity" as "notice or attention given to someone or something by the media," whereas it describes "celebrity" as "the state of being well known."[14] To be clear, celebrity is more than publicity. Although most science and technology enjoy publicity from time to time via media headlines, article features, or special interviews, not all science falls subject to celebrity. In other words, all celebrity involves publicity, but not all publicity leads to celebrity.

Second, in using this new term of "celebrity science," I am not referring to celebrity on the individual level by pointing to the interactions that a single scientist has with the media or the attention they receive.[15] Rather, I am referring to celebrity on the group level and the interest that exists around a subject of science, with a view to how such interest affects the overall group of practitioners associated with it.[16] In a celebrity science, it is the subject of science that is marketed as a commodity to be consumed by the public through the collective promotional activities of the media, research practitioners, and their respective institutions. In this case, for example, it is the celebrity that surrounds the practice of recovering DNA from fossils, its application to questions in evolutionary biology, and even the idea of potentially using it to bring back extinct creatures as embodied in the book and movie *Jurassic Park*.

Third, the celebrity science concept is intended to give a name to what is, in my view, a positive phenomenon. I am aware that "celebrity" carries historical and sociological baggage. It can denote fame and fortune but can also have negative connotations of vanity, superficiality, and inauthenticity. Indeed, celebrity is often used in a pejorative sense. Given this, the celebrity science concept runs the risk of being misinterpreted or reinterpreted across different audiences, and it may alienate some scientists for which this history was written. I hope it will not as I do not intend for the concept to be an unfavorable one.

Finally, I realize that in writing a narrative of ancient DNA research as a case study of celebrity science, I am further contributing to the making of a celebrity science. Doing so will reinforce the media attention in a way that will likely affect the researchers working in or around this practice. In addition, writing about celebrity science also makes me a product of it. There are consequences from situating my work in this context, both positive and negative. Nonetheless, the narrative argument of this book is not only timely but necessary, as media and celebrity culture are increasingly prevalent forces in society today. By situating ancient DNA research in the science communication movement and other shifts toward what science studies scholars call the "mediatization," "medialization," and "celebrification" of science, I show how its history is a product of these phenomena, and how it captures the effects of the ever tighter connection between science and the media.[17]

CHAPTER ONE

Before *Jurassic Park*

DINOSAUR CAPSULE

Charles Pellegrino was a polymath. He was a writer, scientist, and futurist who spread his interests across different disciplines from paleontology and archeology to space science and religion. He helped design rockets and worked in part on the excavations of Pompeii and the *Titanic*. Over the years, he wrote more than a dozen books—fiction and nonfiction alike—and enjoyed close connections with other futurist writers from Isaac Asimov to Arthur C. Clarke. In other words, Pellegrino's interests were wide-ranging, his ideas visionary. He was far from conventional, operating on the fringes of science but often at the center of controversy, especially when it came to his ideas of using multimillion-year-old DNA to bring dinosaurs back to life.

Born in New York in 1953, Pellegrino moved to New Zealand to study at Victoria University of Wellington. He claimed to have earned his doctorate in paleobiology in 1982, but the university denied he ever received the degree. Reportedly, Pellegrino and his committee had a falling-out over evolutionary theory, but his credentials were only questioned decades later when his book on World War II survivors of the Hiroshima atomic bombing was pulled from publication on accusations of fraudulent source material.[1] Bitter conflict at worst, and bad luck at best, have followed Pellegrino throughout his life in an unfortunate series of events, as his works were largely overlooked or openly contested.

One of Pellegrino's most fascinating but neglected pieces of writing was a thought experiment turned article, published in 1985, on how to resurrect a dinosaur. Pellegrino believed that amber specimens—often whole organisms trapped in sticky tree resin then later encased and preserved in a hardened amber capsule—were the ultimate means to rediscover and recreate past life that had become extinct long before humans evolved on earth. He imagined a time when someone, somewhere, would discover the remains of insects pristinely preserved in amber that had once lived more than 65 million years ago, during the heyday of the dinosaurs. In this article, Pellegrino suggested that "ancient bacteria on and in the flies may still be capable of reproducing themselves," and that in their "stomachs may be some undigested bits of their last meals, meals that came from animals, including dinosaurs that roamed the earth millions of years ago." He further imagined the possibility of recovering genetic material from these amber fossils. The DNA would no doubt be damaged and degraded from the millennia, but he proposed using genetic material from living organisms today to fill in the gaps. He then suggested inserting this genetic information "into a cell nucleus" with "a yolk and an eggshell" in order to "hatch" a "dinosaur." If such a specimen existed, scientists could "re-create prehistoric animals."[2] According to Pellegrino, dinosaur resurrection, in theory, could be possible.

Pellegrino claimed his ideas were inspired by two specific events from several years earlier. The first occurred in 1977 when Gerard Case, a local fossil hunter, introduced Pellegrino to a treasure trove of ancient amber from a site in New Jersey dating to the Cretaceous Period, the last and longest span of time in the Mesozoic Era. According to Pellegrino, two digs and two years later, they had recovered two 95-million-year-old flies preserved in amber. The second event took place shortly following his search for amber in the field, when Pellegrino, then at the Smithsonian Institution in Washington, D.C., and entomologist Paul Wygodzinski at the American Museum of Natural History in New York made an unusual observation in the lab. Under the microscope, they found "mummified" insects in amber whose "internal organs appeared to be preserved in microscopic detail" just "as if they have been alive the day before." These two events together prompted Pellegrino to imagine the potential of amber for studying ancient and extinct organisms.[3] Pellegrino speculated that if cellular structures could stand the test of time, then perhaps molecular components like DNA could too. If so, it could be a way of bringing dinosaurs back to life to "study them face to face."[4] Inspired by amber-embedded insects in the field and in the lab, Pellegrino actively advocated for these ideas about the theoretical preservation, potential extraction, and hypothetical resurrection of extinct life.

Pellegrino's ideas, however, faced stiff criticism and resistance. In a later memoir, he remembered having open discussions about his theories with colleagues only to discover they thought them too speculative or "totally bizarre" and even "downright crazy."[5] Even as he experienced difficulty convincing his colleagues of his ideas in casual conversation, Pellegrino—unsurprisingly—encountered even more difficulty publishing them professionally. Based on a series of unpublished correspondence, it appears that Pellegrino had communicated with *Smithsonian Magazine* since 1981 about his amber article, but editors and reviewers alike thought the article was too speculative and therefore unpublishable. John Wiley, the magazine editor at the time, wrote a letter to Pellegrino explaining that even reasonable speculation would be difficult to defend because the magazine tended to be traditional and confronted enough challenges publishing mainstream articles.[6]

There were two reasons for this resistance to Pellegrino's ideas. The first was theoretical. The general scientific consensus at the time was that organic components like soft tissue structures and molecules (amino acids, proteins, nucleic acids) did not preserve in the fossil record.[7] Instead, when an organism died, scientists presumed that organic components decayed over time due to chemical and environmental processes. Internal processes such as autolysis contributed to the self-destruction of an organism's cells soon after death. Taphonomic processes also played a role in an organism's rate of degradation and potential for fossilization. For instance, external factors such as exposure to water and high or low temperatures could certainly contribute to the breakdown of cells and molecules. With these effects combined, often little was left of an organism except its skeleton. However, some studies as early as the 1950s began to challenge existing conceptions of cellular and molecular preservation.[8] A number of studies, for example, actually found evidence that amino acids and proteins could survive millions of years after an organism had died. Important as these findings were, there was no evidence that nucleic acids (DNA) were capable of such long-term preservation. This presented a problem for Pellegrino's hypothesis. Nonetheless, he was confident that amber could protect an organism from the standard chemical and environmental processes of degradation, thus preserving its DNA despite the test of time.

Second, Pellegrino's ideas faced resistance for technical reasons. Even if DNA could be preserved in exceptional environments such as amber, the available technology for extracting and sequencing what would very likely be degraded DNA was inadequate for the job. In the early 1970s, less than a decade before Pellegrino started openly theorizing about his ideas, researchers developed some of the first molecular cloning methods that allowed the direct isolation, replication, and manipulation of DNA in ways not previously possible.[9] A few years later, researchers invented the first molecular sequencing technique for determining the order, or sequence, of nucleotides (adenine, guanine, cytosine, and thymine) in DNA.[10] This technique, called Sanger sequencing, was quickly adopted by researchers worldwide for its accuracy and efficiency, revolutionizing biomedical and biotechnological research along the way. Indeed, the Sanger sequencing was so influential that one of the researchers behind its invention, Frederick Sanger, was awarded the Nobel Prize in Chemistry in 1980 for its invention.[11] Although revolutionary for the time, the technique was most appropriate as applied to modern material where much longer strands of DNA were readily available for cloning and sequencing. If any DNA were ever discovered to be preserved in fossils, it would likely be damaged and fragmented, which would render the cloning and sequencing of such short strands a challenge.

Pellegrino knew the search for and study of DNA from fossils depended on serious technical improvements, but he was confident that with advances, and a lot of luck, researchers would one day recover DNA from the stomachs of ancient insects and bring dinosaurs back to life. Thus, Pellegrino took preemptive steps, placing his prized 95-million-year-old amber insects away in a refrigerator where they would stay safely stored for twenty years or more while he waited for technological developments. Pellegrino certainly did not disregard the necessity of technology, but he also did not let its absence obstruct his vision for what might be possible in the future. For Pellegrino, this was a future "on the verge of redefining the word *extinct*."[12]

Finding difficulty convincing colleagues and publishing professionally, Pellegrino took his recipe for resurrection in a different direction. Eventually, *Omni*—a popular science-fiction magazine—accepted his article on how to bring a dinosaur back to life. And in 1985, "Dinosaur Capsule" was finally published.

THE EXTINCT DNA STUDY GROUP

Pellegrino was not the only one thinking about recovering DNA from fossils. In the summer of 1980, John Tkach—a dermatologist in Bozeman, Montana—was riding his bicycle to work, apparently pondering the extinction of the dinosaurs.[13] Tkach had always been interested in immunology, and he was fascinated by dinosaurs. He, along with other scientists of course, wondered why dinosaurs, apart from the line leading to living birds today, went extinct nearly 65 million years ago. It was a perennial question in the field of paleontology, and while new hypotheses about dinosaur evolution and extinction had begun to circulate a decade before, Tkach found them unsatisfying.

Riding his bicycle to work that summer, Tkach had a thought, an answer to the mystery of dinosaurian extinction. According to a private memoir by Tkach, he hypothesized that dinosaurs likely lacked a "bursa of Fabricius"—a vital organ in immune systems of modern birds. Tkach speculated that dinosaurs, lacking this organ, "could not cope with infections," and that "new pathogens must have come along that they could not defend against," so "over millions of years" they must have "died out."[14] In formulating this idea, Tkach was also aligning part of his hypothesis with an already hotly contested topic at the time, namely that extant, or living, birds today are the direct descendants of extinct dinosaurs. Tkach's idea was controversial to begin with but even more so because it lacked evidence. He would need DNA, specifically dinosaur DNA, as evidence.

That autumn, Tkach had an idea. He imagined a scenario in which "a mosquito had fed on a dinosaur," leaving behind "a dinosaur white blood cell with a diploid set of chromosomes in its stomach." If that "insect had been preserved in amber," he thought, "it might be possible to recover the chromosomes from that white blood cell, put them into an enucleated amphibian egg, and grow a dinosaur." Tkach thought it "unlikely, but theoretically possible."[15] He shared this speculation with several scientists, hoping to enlist their efforts in the search for DNA from ancient amber. In the process, he encountered a mix of pessimism and optimism. Despite mixed reviews, Tkach submitted his ideas to *Developmental and Comparative Immunology,* but his manuscript was rejected on account that his hypothesis had no evidence to support it.[16] Tkach felt his ideas had reached an impasse.

In 1980, the same year Tkach started speculating about dinosaur extinction and resurrection, George Poinar, an entomologist, and Roberta Hess (later Roberta Poinar), an electron microscopist, at the University of California, Berkeley, went to work for what they thought would be another average day in the lab. According to their memoir, this day turned out to be different. Under the microscope, they were shocked to see the insides of a 40-million-year-old insect preserved in amber: "When we looked up at each other, the same thought was written across our faces—surely this fly's cell structure must be intact too!" Together, they began to investigate what they believed was a unique discovery. They first photographed the fossil for documentation, then sliced it in two. Next, they undertook the tedious process of sectioning the insect inside the amber for a more detailed examination. After weeks of work and trial and error, the results were ready. Poinar and Hess found exceptional evidence for the organic preservation of a 40-million-year-old fly in Baltic amber, complete with "nuclei and organelle" and "entire muscle bands with easily identifiable components such as fibrils and mitochondria." They even saw evidence for the "tracheoles, the breathing apparatus of insects."[17]

As far as Poinar and Hess were concerned, this was a rare case of mummification from the prehistoric past. Although they did not find evidence for DNA, they did privately speculate about its likelihood. "If tissues could be discovered so well preserved in amber-embedded insects 40 million years old," they wrote in their memoir, "what else could be found?" "What about nucleic acids?"[18] If DNA preservation were remotely possible, they believed it would be so through the preservative properties of amber resin.

In 1982, *Science*—one of the oldest, most widely read, and most prestigious academic research journals—published Poinar and Hess's research results as the first evidence that soft tissue structures could remain nearly perfectly intact millions of years after an organism's death.[19] Tkach encountered this paper shortly after his own article was rejected, reviving his interest in dinosaur resurrection, particularly from DNA found in amber-preserved insects. In his private memoir, Tkach recalled Poinar and Hess's publication as "pivotal" because it suggested that his own ideas were not so far-fetched but in fact quite "practical."[20] In fact, "it was damned exciting."[21] For Tkach, Poinar and Hess's article on cellular preservation in ancient amber, and principally its publication in a highly esteemed journal such as *Science*, was the empirical evidence he needed. Tkach wrote Poinar in December 1982, and in January of the next year he received a reply. Poinar suggested they form a study group.

In 1983, Tkach, Poinar, Hess, and fellow colleagues formed the Extinct DNA Study Group. By February of that year, Tkach had written and distributed the first "Extinct DNA Newsletter" to a handful of scientists across the country. In the introduction, he stated the new group's principal objective: to investigate the recovery of genes from extinct organisms and their transcription and translation via recombinant DNA technology. The group was also concerned with studying the evolution of proteins as well as the role of parasites, pathogens, and bacteria in the extinction of species. The Extinct DNA Study Group was likewise interested in "culturing tissues from extinct life forms" and "cloning extinct life forms by the recovery of haploid or diploid sets of chromosomes." Overall, Tkach described the Extinct DNA Study Group as a group of "mature scientists" with "extensive educations" and "a hardcore molecular biology outlook."[22] According to the group, the answers to some of the most interesting questions in paleontology and evolutionary biology would be found using molecular biological techniques and technologies.

Following the first newsletter came the first meeting of the group, which Tkach held at his home in Montana in March 1983, and they discussed issues of authenticity and contamination. If they wanted to convince doubtful colleagues of the long-term preservation of DNA in ancient and extinct organisms, they would need to demonstrate that DNA or other molecular information extracted from fossil material was ancient and authentic to the organism of interest and not contamination from other organisms or the external environment. They also discussed the importance of terminology, from "paleobiology" to "paleogenes" and "paleogenomes," as they entered what they felt was unfamiliar territory.[23]

During this first meeting, Tkach shared his hypothesis about using DNA preserved in the stomach of an insect preserved in amber to resurrect dinosaurs. He also outlined the idea in writing, as well as its reception by Extinct DNA Study Group members, in the second newsletter.[24] Reportedly, the group was intrigued but really doubted DNA could remain sufficiently intact to be viable.[25] From somewhat isolated speculation to more collegial collaboration, the Extinct DNA Study Group became the first official forum through which researchers entertained ideas regarding the preservation and extraction of DNA in fossils and resurrection of extinct species.

The group knew their unconventional ideas would generate criticism. Sure enough, they encountered this when recruiting researchers to join. In a later memoir written by Poinar and Hess, they recalled that "few were willing to jeopardize their careers or suffer the derision of colleagues." As far as they were concerned, "the actual membership was limited to a courageous handful."[26] But even the "courageous" were cautious. In his own account, Tkach recalled that "all members of the group feared for their reputations" because "they were concerned that premature release of information without adequate proof of authenticity" would "damage their reputations." "We agreed to be careful about our claims," he recounted, "until we felt we had done as much as we could to prove authenticity."[27] According to the first newsletter, Tkach expressed this concern, writing, "Dr. Poinar has asked that we not discuss this work with the press" for the reason that "adverse publicity could have a crushing effect on his work."[28] From their view as practicing researchers, speculation was a creative and healthy part of the scientific process. However, too much speculation with too little evidence, coupled with premature exposure, could damage the research and reputations of those involved. Timing and evidence were everything.

Poinar and Hess had at first assumed they were alone in the search for cells and molecules in multimillion-year-old amber insects. So did Tkach. Their intellectual isolation, however, was short lived. "It's amazing when you consider it," claimed Poinar and Hess, "how people can arrive at the same idea (extracting DNA from amber insects) from different perspectives."[29] Tkach's interest in dinosaur extinction and Poinar and Hess's entomological investigations in the lab brought them to an identical conjecture about how to travel back in time to learn more about the prehistoric past, and even bring dinosaurs back to life.

JURASSIC PARK

Around this same time in the early 1980s, Michael Crichton—a doctor turned novelist—was writing a screenplay about a genetically engineered dinosaur. A graduate of Harvard Medical School, Crichton had foregone practicing medicine in exchange for a career writing science fiction. Over the course of a decade, he had earned a name for himself as a world-renowned author whose best-selling novels *The Andromeda Strain* and *Westworld* were made into highly successful movies. Crichton wanted to write another science-fiction thriller, this time about dinosaurs, but he feared the book would appear too trendy given what he referred to as an "enormous mania about dinosaurs."[30]

Sure enough, the previous decade had witnessed a growing professional and popular interest in dinosaurs. This "Dinosaur Renaissance" was stimulated by an influx of research that questioned traditional views of dinosaur anatomy, physiology, evolution, and extinction.[31] Some paleontologists, for example, argued that although most dinosaurs went extinct, one specific lineage had survived and evolved into the feathered, flight-capable birds of today.[32] Others argued that dinosaurs had been hot-blooded creatures, not cold-blooded reptiles as previously presumed and as often depicted in media.[33] Additionally, new fossil evidence of dinosaur skeletons with associated nests and eggs led scientists to suggest that these ancient animals likely cared for their young.[34] Perhaps most revolutionary, however, was a new hypothesis that proposed a massive asteroid impact on earth as the chief cause for the extinction of nearly 80 percent of life on the planet, including dinosaurs, approximately 65 million years ago.[35]

The public had long been captivated by prehistoric life, dinosaurs in particular, thanks to early scientists' efforts to popularize the young field of paleontology in the late nineteenth century and throughout the twentieth. The "Dinosaur Renaissance" of the 1970s was very much a continuation of this obsession with prehistory.[36] Crichton realized this, along with the fact that the dinosaur mania was not fleeting. As far as he was concerned, this "fascination with dinosaurs was permanent," and he decided to capitalize on it.[37]

As Crichton was writing his story, Poinar and Hess were publishing their research on the long-term cellular preservation of insects in amber. In a later account of the event, Poinar told a media reporter and blogger that Crichton visited him at Berkeley back in 1983. According to Poinar, Crichton took notes while he discussed the details of his work. As far as Poinar could remember, Crichton never mentioned anything concerning writing a book about genetically engineered dinosaurs.[38] After this meeting, Poinar and Hess went on with their efforts, particularly their collaboration with the Extinct DNA Study Group, not thinking of Crichton until years later.[39]

Based on Poinar and Hess's report of events, the next time Crichton crossed their minds was when they received a phone call from Universal Pictures informing them that they, as well as the Extinct DNA Study Group, were acknowledged in the back of a new book titled *Jurassic Park*.[40] Initially published in November 1990, the science-fiction thriller was about genetically engineered dinosaurs that scientists had resurrected via DNA preserved in and extracted from a long-dead mosquito entombed in amber. These fictional scientists intended the dinosaurs to be part of a world-class theme park, called Jurassic Park. However, it quickly turned into a scientific experiment gone madly wrong. In the book, Crichton very intentionally tapped into the promise and fears of genetic engineering, appealing to professional and public audiences alike.[41] Jurassic Park was an immediate best-seller and a multimillion-dollar movie in the making by Universal, directed by Steven Spielberg, the critically and commercially acclaimed director of the Hollywood blockbusters *Jaws* and *E.T. the Extra-Terrestrial*.[42] At the end of the book, Crichton acknowledged the ideas and individuals that inspired it. Just as the caller had said, Poinar, Hess, and the Extinct DNA Study Group were clear contributors.

Although Poinar and Hess's research was acknowledged, Pellegrino's ideas as recorded in "Dinosaur Capsule" and published by *Omni* in 1985 were never mentioned, at least not at first.[43] In one version of the book's acknowledgments printed in 1991, Crichton noted that "certain ideas" about "paleo-DNA" were "first articulated by George O. Poinar, Jr., and Roberta Hess, who formed the Extinct DNA Study Group at Berkeley."[44] In another paperback version published the same year, the acknowledgments read slightly differently. Here, Crichton noted that "certain ideas" were "first articulated by Charles Pellegrino" but "based on the research by George O. Poinar, Jr., and Roberta Hess, who formed the Extinct DNA Study Group at Berkeley."[45]

To this day, it is unclear as to what role Pellegrino's "Dinosaur Capsule" played, or did not play, as inspiration for Crichton's *Jurassic Park*. What is clear, however, is that Pellegrino felt cheated, not necessarily by Crichton but certainly by Poinar. Recall that Pellegrino had corresponded with *Smithsonian Magazine* nearly a decade before *Jurassic Park* was published about his hypothesis to resurrect dinosaurs from DNA preserved in amber insects. A letter from John Wiley, the magazine editor, to Pellegrino had explained that reviewers thought his article was too speculative. The letter also revealed that Poinar had been one of those reviewers and an obstacle in its path toward publication. According to Wiley's letter, Poinar seemed to want to write and take credit for the idea himself.[46]

At the time, Pellegrino knew Poinar was aware of his work and for whatever reason was opposed to its publication. However, it was only after *Jurassic Park* was released and the acknowledgments as well as media reporters began to credit Poinar as the inspiration behind the book that Pellegrino became openly upset about the situation. Indeed, Poinar now seemed to be taking credit for the idea he had once rejected. An article in the *New York Times*, for example, clearly attributed the "recipe" to Poinar. "Obviously, we couldn't reconstruct an extinct animal today, even if we had all its DNA," said Poinar in an interview with the *Times*. "However, my belief is that there are dinosaur cells inside biting flies trapped in amber of Cretaceous age and older. It's just a matter of finding the dinosaur DNA and getting it out."[47] In response, Pellegrino wrote the newspaper asking that they set the record straight. "There is an old rule in science, about the reception of a new idea," he observed, referencing one of Arthur C. Clarke's famous sayings. "At first it is dismissed as being wrong, then the one who conceived it is character-ized as being a heretic, a lunatic, or both; and finally, it is said to be some-thing everybody knew all along."[48] From Pellegrino's view, Poinar took this a step further, and a step too far, so much so that Pellegrino threatened Poinar with legal action for misappropriation of his work, namely his idea that scientists might one day discover dinosaur DNA from amber-preserved insects and bring them back to life.[49]

UNLIKELY BEGINNINGS

Starting in the late 1970s and continuing throughout the 1980s, a number of characters from futurists and enthusiasts to scientists and science-fiction writers contributed to the early intellectual history of the search for DNA from fossils. At this time, three ideas—the theoretical preservation and potential extraction of DNA from fossils and the hypothetical resurrection of extinct organisms from DNA in ancient material—inspired a new way of thinking about fossils. These ideas appeared to have arisen independently among different individuals from four distinct perspectives. Pellegrino, for example, was a scientist and futurist who approached these ideas as a visionary, so much so that scientific colleagues were hesitant or sometimes downright dismissive of his resurrection hypothesis. Tkach, a science enthu-siast, was similarly speculative but ultimately successful in enlisting experts to investigate his hypothesis, while Poinar and Hess appeared to have arrived at these ideas through more conventional means in the lab and publication of research results in the high-profile journal *Science*. Crichton's fictional work in progress was also a part of this early history. These story lines suggest that scientific innovation can originate from ideas and indi-viduals outside the traditional confines of the research laboratory.

One way to understand the emergence of these unique but not unrelated events is to view "Ancient DNA" as a "boundary object." This concept of a boundary object was proposed by science studies scholars Susan Leigh Star and James R. Griesemer as a framework for understanding how information or material can be used in various ways by various people. As they argued, a boundary object is flexible enough to be interpreted differently across groups but stable enough to maintain a recognizable identity across these same groups. For example, in examining the founding of the Berkeley Museum of Vertebrate Zoology in the early 1900s, Star and Griesemer argued that maps, field notes, and even specimens functioned as boundary objects as multiple people from museum curators and scientists to amateur collectors and trappers viewed these objects differently and used them to different ends.[50] Ancient DNA was a kind of boundary object but not, at least at this time, in a material or tangible sense. Rather, it was a boundary object that existed as an idea. It was the idea of isolating, extracting, and sequencing DNA from fossils and using it to study evolutionary history (or possibly using it to bring back extinct creatures such as dinosaurs) that disparate people, arguably from disparate social worlds, had in common. The potential to discover ancient DNA functioned as an organizing principle, bringing different people together to investigate its reality, albeit to varying ends.

These individuals were instrumental in setting expectations in terms of what they imagined researchers could be capable of achieving with this new line of research. Early expectations regarding the recovery of DNA from fossils and its potential uses within evolutionary biology were not isolated imaginings of a single person or group. In other words, various individuals—other than the actual scientists directly involved in the research—often have their own visions for scientific and technological innovation, as well as their own opinions regarding the application of those innovations. Equally important is understanding the role that these expectations, however grounded or fantastical, may play in the early days of innovation, particularly the ways in which expectations can generate activity around new ideas.[51] For Tkach, the recovery of dinosaur DNA from amber insects could be used to test his hypothesis of dinosaurian evolution and extinction, ultimately leading to a connection with Poinar and formation of the Extinct DNA Study Group. For Crichton, it was a fantastical but plausible idea that informed the premise for his novel.

From its inception, the idea of extracting DNA from ancient and extinct organisms was closely connected to the idea of bringing dinosaurs back to life, playing out in the work and writings of various individuals—scientists and visionaries alike—well before the book and movie *Jurassic Park* made its premiere. More than that, this connection would prove foundational to the birth and rise of a new scientific field.

CHAPTER TWO

Ideas to Experiments

DISCOVERING ANCIENT DNA

Ideas about the theoretical preservation and potential extraction of DNA from fossils originally arose outside the traditional laboratory research setting, but it was at the University of California, Berkeley, that ancient DNA research became a localized, specialized, and more widely recognized activity. Allan C. Wilson, a professor of biochemistry at Berkeley, was primarily responsible for transforming the idea of recovering DNA from fossils into experiments with evidence for the long-term preservation of molecules. Born in New Zealand in 1934, Wilson moved to the United States for his graduate studies, ultimately completing his Ph.D. at Berkeley in 1961, where he would spend the rest of his career. Over the years, Wilson earned a reputation as a pioneering scientist in evolutionary biology for his use of molecular data to reconstruct patterns and processes in evolutionary history.[1]

Specifically, Wilson was interested in using molecular biological techniques to study human evolution. One of his most well-known studies, published by *Science* in 1967, forced the scientific community to reconsider the way human evolutionary history was understood.[2] This article, co-authored with his graduate student Vincent M. Sarich, was considered a landmark paper in the field for its use of molecular evidence from humans and apes to suggest a much more recent common ancestry than previously thought based on paleontologists' estimates using fossil evidence. In this paper, Wilson and Sarich relied on a comparison of protein sequences to determine that humans and primates split from one another approximately 4–5 million years ago.

While Wilson and Sarich's conclusion was groundbreaking, so was their methodology. At the time, their study lent support to a new and highly controversial hypothesis—the molecular clock hypothesis. This concept—introduced by Linus Pauling, a physical chemist at the California Institute of Technology, and his postdoctoral student Émile Zuckerkandl—suggested that molecules evolve over time at a steady rate, and because of that, they proposed using molecular data and their mutation rates to date divergences of one species from another in their evolutionary history.[3] As historians of science Marianne Sommer and Elsbeth Bösl both note, the molecular clock hypothesis essentially argues that history, specifically evolutionary history, is recorded in molecules, written in our DNA.[4] Zuckerkandl and Pauling suggested this concept in general, but it was Wilson and Sarich who demonstrated evidence of it as applied to the evolutionary history of primates and humans.

Wilson had long sought to apply molecular biological techniques and data to study evolutionary history, and he soon thought to extend this research to ancient organisms too. In fact, doing so appeared a natural extension of his expertise and visionary tendency. Poinar and Hess's publication in *Science* in 1982 on the exceptional cellular preservation of a 40-million-year-old insect in amber garnered widespread attention from many, including Wilson. Whether Wilson contacted Poinar, or whether Poinar contacted Wilson, is not entirely clear.[5] What is clear, however, is that Wilson's lab and Poinar's lab—both at Berkeley at the time—eventually established a collaboration.[6]

In 1983, Poinar and Hess, together with Russell Higuchi—a molecular biologist and postdoctoral researcher working with Wilson—set out on the first experiment to actually test ideas about the preservation and extraction of DNA from insects in amber. The challenge was clear, as preparation alone was a tedious task. First, Poinar began by selecting amber specimens that would potentially offer optimal preservation of DNA. Next, the three sterilized all the lab equipment in order to avoid contamination. Even as early as the 1980s, they all realized the risk of contaminating their results with their own DNA merely through handling the specimen during preparation and extraction. Therefore, they took precautions to minimize contamination, thus maximizing the chances of recovering DNA from the organism itself. Meanwhile, Hess began to section and take samples of the specimens by cutting into their amber capsules, then carefully removing the insects' tiny tissues.[7]

After specimen preparation was completed, they performed a template assay test on the tissues, an experiment that made a radioactive copy and gave a radioactive signal in the presence of even the most miniscule amounts of DNA. If DNA were present, it would be copied and a signal would be emitted. In the end, they found evidence of DNA in two of the seven specimens sampled—a moth and a fly. They appeared to overcome the biggest challenge, how to get DNA from multimillion-year-old amber insects, but another issue soon emerged, namely how to determine to whom or what the DNA belonged. Was the DNA from the insect itself or was it a contaminant from the environment? At the time, no other experiments were done to determine the authenticity of the DNA, and the results were never published.[8]

Following this initial experiment to recover DNA from amber insects, Higuchi and Wilson shifted their efforts to a much younger specimen, the extinct quagga that once roamed the plains of South Africa over a hundred years ago. *Equus quagga* was an enigmatic species primarily because of its unusual appearance: its backside and hind limbs were solid brown, resembling a horse, while its neck and face were covered by brown and white stripes, resembling a zebra. By the late 1800s, however, the entire species had become extinct. In fact, humans had hunted the quagga to extinction, leaving behind the skeletons and skins of only twenty specimens, now stored in museums across the world.

There were a number of reasons behind Higuchi and Wilson's decision to study the quagga. The first was born out of a sentimental motivation. Reinhold Rau, a taxidermist and conservationist at the South African Museum in Cape Town, near where quagga once lived a hundred years ago, felt a sense of guilt about the role humans played in the species' demise. Indeed, Rau was interested in bringing the quagga back to life in order to rectify its extinction. His plan was to use selective breeding processes with living zebras to recreate the quagga's unique striping pattern. Although Rau did not plan to use DNA to try to clone the quagga, he did need its DNA to determine its phylogenetic relatedness to extant horses and zebras so his breeding program could be successful.[9] During his career, Rau visited several museums looking for suitable samples for DNA analysis. He ultimately took a soft tissue sample from a

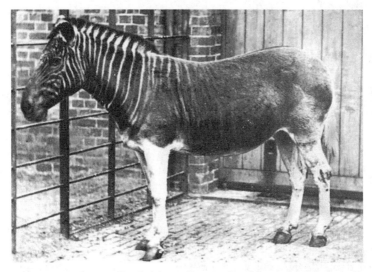

Equus quagga. This mare lived at the London Zoo from 1851 to 1872. (Reprinted by permission from Springer Nature: *Nature*'s News and Views, "Raising the Dead and Buried" by Alec Jeffreys, copyright 1984)

140-year-old quagga specimen on display at the Natural History Museum in Mainz, Germany. Rau sent the sample to Oliver Ryder at the San Diego Zoo, an expert in molecular evolution and conservation. It was Ryder who offered the sample to Wilson and Higuchi at Berkeley for DNA analysis.[10]

While there were sentimental motivations for studying the quagga, there was also a scientific interest in its evolutionary history. At the time, paleontologists who used morphological data, such as fossil data, disagreed over the quagga's relatedness to living horses and zebras. Some argued it was more closely related to horses than zebras. Meanwhile, others argued it was more closely related to zebras but constituted a separate species. Some disagreed with both interpretations, proposing that the quagga was not a separate species of zebra but a subspecies of the Plains zebra.[11]

Finally, there were practical reasons for studying the quagga. Sure enough, Higuchi and Wilson thought it would be much more likely to recover DNA from material a few hundred years old than trying to gather DNA from material millions of years old. And if they were successful in doing so, they could much more easily verify the authenticity of the extinct quagga's DNA, as well as determine its evolutionary history, by comparing it to DNA sequences of extant horses and zebras of today.

In the spring of 1984, Higuchi, Wilson, and their team of researchers successfully extracted DNA from the 140-year-old remains of a quagga. In taking a sample from a small piece of dried muscle, Higuchi and colleagues were able to recover sequences of mitochondrial DNA, a type of DNA inherited on the maternal line and found in abundance in plant and animal cells. To do so, they used standard methods of DNA extraction, but the process was complicated by the fact that the quagga DNA was old, degraded, and fragmented. In fact, they were only able to recover approximately 1 percent of that expected from fresh muscle. In order to better study these DNA fragments, Higuchi and colleagues cloned and amplified single fragments of the DNA using a phage vector—a type of DNA molecule that naturally replicates itself in bacteria. Just as clones of a single human, if they existed, would share the same DNA sequence, clones of a single fragment of quagga DNA would also have identical DNA. Of all the clones made—nearly 25,000 total—just two could be demonstrated to be mitochondrial in origin and were subsequently sequenced. The two clones together included 229 nucleotide base pairs of quagga DNA. Although short strands, the DNA sequences were similar enough and yet different enough from zebra, horse, cow, and human mitochondrial DNA sequences to show that they had successfully obtained, for the first time, ancient DNA from an extinct creature.[12] At the same time, the sequences were most similar to the plains zebra of all the zebras tested, and least similar to the horse of all the equine species tested. This was evidence for including the quagga as a subspecies of the plains zebra. With this molecular data, they were also able to determine that the quagga had diverged from the zebra approximately 3–4 million years ago.

Higuchi and colleagues took a number of steps to announce their discovery. First, they wrote up their research results and submitted an article for publication in *Nature,* a renowned British journal and century-old competitor with the American journal *Science.* While they waited for their paper to be reviewed, Higuchi and colleagues worked quickly to write and submit a grant proposal to the National Science Foundation (NSF), one of the largest federal government agencies responsible for funding scientific research across the United States. The application they submitted— "Molecular Paleontology: Search for Fossil DNA"—requested $330,000 to be distributed over three years and represented the first official research proposal of its kind. Wilson and Higuchi would head the research with Poinar, Hess, Alice Taylor (an electron microscopist), and Barbara Bowman (a graduate student who worked on the quagga study). Their goal was to continue the search for DNA from the quagga and other extinct species including bison, mammoths, and moas—an extinct and flightless bird of New Zealand. They even proposed to continue searching for multimillion-year-old DNA from amber insects.

In their application, Wilson and colleagues stated that this research, if funded, could mark the start of a new way to study evolutionary history and give rise to a new field of scientific inquiry, molecular paleontology. "This is the first proposal to study the possible utility of DNA to paleontology," they wrote. "If clonable DNA is present in many fossil bones and teeth and in insects included in amber, a new field, molecular paleontology, can arise." However, the quagga data alone formed both the preliminary and primary evidence on which the proposal stood.[13] The application was a bold one, especially given the available evidence on which it rested. Indeed, they conceded in their application the "exotic" and "speculative" nature of their project, but they remained optimistic that their work could be a revolutionary approach to studying ancient and extinct life. They had high hopes for this new venture, but their proposal was in the hands of other scientists and a decision panel who would either accept or reject it.

NSF reviewers were intrigued by the prospect of searching for DNA from fossil material, but caution colored their feedback. One reviewer, for example, called the proposal "interesting, significant, even exciting." Another called it a "pioneering effort" at the interface of "molecular systematics and paleontology." Meanwhile, another reviewer was not as optimistic: "Discovering and extracting DNA from fossil species is a very interesting and technically difficult biochemical feat, but it is certainly not clear to me how this approach will broaden our perspective on any major evolutionary problems." Other referees recognized the difficulty of the task Wilson and colleagues proposed but were not as quick to give up on it. Sure enough, they felt the challenge was worth the effort. "I refuse to gaze into a crystal-ball and reject the possibility *a priori*," wrote a reviewer. "It is clear that *looking* for fossil DNA is worth the trials and tribulations, particularly if so distinguished [a] researcher as Wilson wishes to undergo the trauma." A different reviewer made a similar comment: "I am not convinced that selection of these organisms will demonstrate the universal applicability of recombinant DNA technology to systematic evolutionary studies. . . . However, at one time it was common knowledge that the earth was flat and the moon was made of green cheese." Despite all the technical obstacles, a handful of reviewers seemed confident that Wilson and colleagues would overcome them. "If it is possible to do," wrote one last reviewer, "they can do it."[14]

Despite the mostly positive reviews, the decision panel unanimously rejected the proposal for funding. Their first reason for rejecting it came down to a matter of tangible research outputs: "The proposal is not designed to develop a new technology which when developed would be broadly applicable to a wide range of specimens." The second reason came down to the practicality of the research itself. "At most the project will provide that some fossil remains contain clonable DNA," explained panel members. "If clonable DNA is obtained, its usefulness for phylogenetic studies remains to be shown, given the likelihood of the occurrence of unquantifiable diagenetic change and the presence of contaminating DNA." Overall, the panel was far from convinced that the search for DNA from fossils would prove a worthwhile endeavor: "The Panel does not consider that obtaining clonable DNA from a 140-year-old museum specimen, however interesting, provides sufficient preliminary evidence that DNA from 10,000 or 26-million-year-old specimens is likely to yield valuable information."[15] With that, the search for DNA from fossils seemed at a standstill.

MUMMY DNA

While Wilson and Higuchi were experimenting with amber fossils and quagga remains, similar studies regarding the long-term preservation of molecules were being pursued elsewhere. Svante Pääbo, a doctoral student at the University of Uppsala in Sweden, had recently begun exploring the idea of recovering DNA from ancient Egyptian mummies. Although a student of molecular biology, Pääbo had always been interested in Egyptology. Indeed, ancient Egyptian mummies and culture had fascinated archeologists, anthropologists, and linguists, as well as scientists and the broader public, for centuries.[16] Pääbo was determined to somehow couple his passion for ancient Egyptian culture with his studies in molecular biology, and his knowledge of current research from prominent scientists in the field inspired him to do just that. Pääbo was well aware of Wilson's work on molecular evolution, and like Wilson, he was attracted to applications of molecular biological techniques to study evolutionary history, especially human evolutionary history.[17] Pääbo also knew of another researcher, Alec J. Jeffreys, a prominent molecular biologist at the University of Leicester in England studying the genetic evolution of humans and apes.[18] According to Pääbo, both Wilson's and Jeffrey's respective works prompted him to speculate on how molecular biological techniques could be used to study ancient life, and in this case to study life in ancient Egypt.[19]

In the summer of 1981, Pääbo went to the store to buy a piece of liver. He was interested in the theoretical preservation of DNA in ancient Egyptian mummies, and he knew that mummification involved dehydration, which would likely prevent DNA degradation. Thus, mummification could prove an ideal process for long-term DNA preservation. Given this reasoning, Pääbo decided to replicate the procedure, to some extent, by cooking the liver in an oven. Pääbo wanted to conduct this work in the lab, but he made sure to keep it secret for fear of being reprimanded by his supervisor or humiliated in front of his colleagues should the experiment turn out to be a total failure. According to Pääbo's memoir, he went to the lab and began baking the liver in an oven, heating it up and drying it out over the course of a few days. By the second day, the smell was so strong and repulsive that he became worried his colleagues, or even worse his supervisor, would investigate and discover the source. Fortunately for Pääbo, as he recalled in his memoir, the smell subsided after another day or two, with no questions asked. In the clear, Pääbo assessed the liver—now a hard, dry, shriveled substance—and attempted to recover DNA from it. According to Pääbo, he was instantly successful. The DNA was certainly fragmented, including just a few hundred nucleotide pairs, much less than the thousands of nucleotide pairs to be expected from fresh tissue. Still, in theory, the study was successful. Pääbo felt "vindicated."[20]

Although this experiment was successful, the preservation and extraction of DNA hundreds to thousands of years old remained to be tested. To satisfy his curiosity, Pääbo asked a curator of a small local museum who also happened to be a close friend if he could take samples from several mummy specimens to try to discover DNA. Although Pääbo was not allowed to take direct samples from some of the most prized and well-preserved mummies, the curator did allow him to sample previously detached or already damaged skin and muscle tissue from three mummies in the collection. Pääbo was appreciative of the opportunity, and back in the lab he applied standard DNA extraction techniques to each of the three samples. To his disappointment, he was unsuccessful in recovering any DNA samples.[21]

Despite this setback, Pääbo was determined not to give up. With help from his museum curator friend, Pääbo approached the Berlin State Museums asking for more and better samples of mummy material. The curators were agreeable, and after a two-week trip to Berlin, he returned to Sweden with more than thirty samples. He continued to work nights and weekends in the lab to keep his study a secret, but this time he decided to investigate each sample for cellular preservation before trying to extract DNA. If there was evidence of cellular preservation, then there might be evidence of DNA preservation. Using a standard technique in microscopy, Pääbo prepared and placed the mummy samples on small glass slides, then stained the samples with a dye. Looking at the slides under a microscope, the dye would emit a color and enhance the visualization of cells or cell parts if any were preserved in the samples. Out of all thirty samples, Pääbo found only three showing even the slightest evidence of cellular preservation. He hoped that at least one of these samples would show signs of DNA preservation too.[22]

Under the microscope, Pääbo further assessed a sample of skin taken from the left leg of a mummified child that had showed very clear evidence for the preservation of cell nuclei. The cell nucleus was particularly interesting because of its role as the command center for cell growth and as a storage unit for the majority of its genetic material, including its DNA. Pääbo applied a second stain to the sample, this time in search of evidence of DNA. According to him, the test was positive, emitting color and therefore evidence that at least some DNA remained intact. Pääbo was elated. As far as he was concerned, this was a sure sign of ancient and authentic DNA. "Since this DNA was in the cell nuclei, where the cellular DNA is stored," he explained in his memoir, "it could not possibly be from bacteria or fungi because such DNA would appear at random in the tissue where the bacteria or fungi were growing." As far as he was concerned, this was "unambiguous evidence" that the DNA observed was that from the mummy child itself and not a contaminant from the environment.[23] Pääbo then attempted to extract the DNA. To his surprise, he was successful.

Pääbo quickly began writing up his research methods and findings for publication. To acknowledge the East German curators who allowed him to sample the specimens, he chose to publish the first recovery of DNA from two-thousand-year-old mummies in an East German journal called *Das Altertum*. His paper was published in 1984, but its reception was anticlimactic. According to Pääbo, this research, which he considered groundbreaking, received no attention at all. The first formal report on the preservation and extraction of DNA from ancient Egyptian mummies received no response—no letter, no question, not even a reprint request. Pääbo, disappointed and discouraged, reasoned the lack of attention was due to publication in a lesser-known journal. "I was excited," he recalled, "but no one else seemed to be."[24]

In an attempt to reach a wider audience and attract greater attention, he sent a second manuscript for review to the *Journal of Archaeological Science*. The journal received the manuscript in October 1984, and although the article was accepted, the review process was slow and the paper was not published until later the following year. Between low readership in the first case and slow publication in the second, Pääbo questioned if anyone cared at all about the prospect of procuring DNA from ancient or extinct material and the implications for paleontology, archeology, and evolutionary biology more broadly.[25]

PUBLISHING ANCIENT DNA

In 1984, the same year Pääbo sent his work out for publication, *Nature* published Higuchi, Wilson, and colleagues' article on the recovery of DNA from the extinct quagga. This piece stood as the first recorded evidence for the long-term preservation of DNA and its successful extraction from the remains of an ancient and extinct species.[26] Their paper and its presentation in an esteemed journal such as *Nature* was noteworthy for three reasons.

First, the quagga study outlined the theoretical and technical procedures for identifying, extracting, amplifying, and sequencing DNA from ancient and extinct material. This was something that had not been previously demonstrated. Second, the study showcased how ancient DNA sequences could be successfully analyzed and applied to phylogenetic problems. It also lent support to the molecular clock hypothesis, an increasingly popular idea that proposed molecules and the rates at which they mutate over time could be used to date the split of one species from another in their evolutionary histories. Finally, the quagga study was a conceptual contribution to the fields of evolutionary biology and molecular biology. Although some of the questions asked and methods used were not necessarily groundbreaking for the time, the fact that Higuchi and colleagues undertook experiments to test ideas about the theoretical preservation and potential extraction of DNA from ancient and extinct specimens made this work significant and its implications enticing. This publication was essential in bringing this new line of research to the attention of the scientific community. Indeed, ancient DNA research could be a way to travel back in time to study evolution in action.

Pääbo was shocked to learn of a successful study so similar to his own, and surprised at its publication in an esteemed journal such as *Nature*. At the same time, he appreciated this work because it validated his own. "If Allan Wilson was studying ancient DNA, and if *Nature* considered an article about 120-year-old DNA interesting enough to publish," wrote Pääbo in a memoir, "then surely what I was doing was neither crazy nor uninteresting."[27] Wilson, an established experimentalist well-known for his molecular work in evolutionary biology, certainly lent credibility to the speculative idea of recovering DNA from fossils. A researcher like Wilson at an institution like Berkeley possessed the authority, security, and resources to test such abstract ideas, while a renowned journal like *Nature* carried the prestige to generate attention and influence over other scientists and the public.

To be clear, publication of the quagga study in *Nature* was effectively an endorsement of its legitimacy. While the evidence in support of the study was important and necessary, it was how the professional scientific community reacted to the evidence—in this case its acceptance and publication—that gave the study the stamp of credibility it needed. This was especially valuable considering the highly speculative nature of the research in the first place. Further, publication in top-tier journals like *Nature* and *Science* carried clout with more than just the scientific community. In fact, the high-profile and high-impact nature of the research these journals published appealed to the wider public. Media reporters were apt to cover them. Given this, scientists worldwide started to think twice about the prospect of DNA hidden away in ancient skins and tissues. In other words, place and power of place mattered to the reception and exploration of speculative ideas.[28]

Inspired, Pääbo wrote a third and final paper on the recovery of DNA from ancient Egyptian mummies, this time submitting it for publication in *Nature*.[29] Lucky for him, it was quickly reviewed and printed in April 1985. In this article, Pääbo argued that DNA could provide genetic answers to historical and archeological questions about Egyptian culture, evolution, population, and disease. However, like Wilson's 1983 amber study and the 1984 quagga study, in Pääbo's study DNA was only partially preserved and only in some samples. In this case, only one mummy exhibited evidence of DNA. If paleontological and archeological specimens were to be reliable data resources for molecular evolution studies, then DNA's survival in specimens would need to be a repeatable, not a rare, occurrence. Ancient DNA would need to be more than an anomaly. Pääbo hoped this publication in *Nature* would finally attract the attention he originally anticipated and encourage further exploration on a much more significant scale.

As Pääbo recalled in his own later account of events, he was excited to see another scientist, especially a scientist as prominent as Wilson, invested in exploring the long-term preservation of DNA: "I thought about how to approach Allan Wilson—a demigod, in my view—to ask if I might work with him at Berkeley after my PhD defense." Indeed, Pääbo was unsure how to introduce himself and his work, so he settled on sending Wilson a copy of his *Nature* paper. Wilson had no knowledge of Pääbo as a researcher, but the manuscript made a considerable impression. Pääbo recalled, "I received a response from Allan Wilson, who addressed me as 'Professor Pääbo'—this was before both the Internet and Google, so there was no obvious way for him to find out who I was." "The rest of the letter was even more amazing. . . . He asked if he could spend his upcoming sabbatical year in 'my' laboratory!" For Pääbo and his lab mates, it was all a "humorous misunderstanding." "I joked with my lab mates," he wrote, "that I would have Allan Wilson, perhaps the most famous molecular evolutionist of the time, wash gel plates for me."[30] Pääbo quickly replied to Wilson to explain that he was not a professor but a student still working on his doctorate. Instead, Pääbo asked if there would be an opportunity after his graduation to join Wilson in Berkeley for his postdoctoral training.

Although ideas regarding the theoretical preservation and potential extraction of DNA from fossils arose outside the traditional laboratory setting, the circulation and reception of those ideas largely depended on evidence produced inside the lab. In other words, the acceptance of these ideas heavily depended on scientists' ability to turn them into experiments with evidence. This was vital, especially considering the idea of discovering DNA from ancient and extinct species was one that many, including researchers themselves, considered wildly speculative. The lab—specifically Wilson's lab in Berkeley and, to a certain extent Pääbo's lab in Uppsala—became the first sites of ancient DNA activity to provide the earliest evidence in support of a new line of research.

RAISING THE DEAD AND BURIED

In the mid-1980s, the search for DNA from fossils, mainly a private affair, went public. Along with the published quagga article, *Nature* also issued a "News and Views" commentary to go with it. The review—titled "Raising the Dead and Buried"—was written by Alec Jeffreys, a respected geneticist, and in it he speculated on the significance of this new line of research. "Is the quagga as dead as a dodo? Not entirely, and nor indeed might be the dodo, if the remarkable findings of Russell Higuchi, Allan Wilson and co-workers . . . are anything to go by." Based on the research findings, the extinct quagga's DNA had survived intact for more than a hundred years, and there was enough of it for scientists to clone and study. For Jeffreys, these findings were certainly preliminary, but they also pointed toward the beginning of something exciting and revolutionary. "Any hopes that molecular biology and paleontology can be fused into a grand evolutionary synthesis by studying fossil DNA, still look like nothing more than a glorious dream," noted Jeffreys. "However, it is far too early to give up, and it might just be possible that DNA has survived in some fossilized material."[31]

For Wilson, the quagga study was his first to exhibit evidence of DNA from an ancient and extinct organism, but he and his lab were no strangers to the study of ancient molecules, nor were they strangers to the press and public attention associated with it. Before both the amber and quagga experiments, Wilson and colleagues had attempted the extraction of proteins and even DNA from woolly mammoths. In the summer of 1977, the opportunity presented itself when an approximately forty-thousand-year-old baby mammoth was found preserved in permafrost near Magadan in Siberia. Named Dima, the baby mammoth was an exceptional fossil find for two reasons. First, it was the most complete mammoth discovered since the 1800s. Second, it was the only complete mammoth to be excavated, then immediately refrigerated in a lab, thus preventing the specimen from thawing and decomposing. Wilson heard about the find and was especially interested for the latter reason, since refrigeration might prevent the degradation of cellular and molecular material. In the spring of the next year, after a series of inquires that eventually resulted in an American-Soviet scientific collaboration (an interesting collaboration for an unusual time of American-Soviet political conflict), a sample of muscle from the carcass was packed in dry ice and shipped to Berkeley from the Soviet Union.[32] Being frozen for thousands of years, Dima the baby mammoth presented a unique opportunity for immunological, chemical, and molecular research of an ancient and extinct species.

Dima's discovery and delivery to Berkeley was of immediate interest to scientists and the public alike, making headlines across multiple newspapers.[33] Indeed, there were good reasons behind all the attention. For starters, humans have long been captivated by mammoths in terms of their existence and extinction approximately ten thousand years ago.[34] For the most part, the mammoth's evolutionary history and its relationship to extant elephants were uncertain, and the reasons for its extinction also remained a mystery.[35] Before Wilson and colleagues even began their work, much less published the results of it, media reporters began to speculate on the implications. Reporter Walter Sullivan for the *New York Times,* for example, conceded that the main goal of the scientists' work was to search for mammoth proteins and perhaps mammoth DNA, with the hope that molecular evidence would shed light on the relationship between the extinct mammoth and extant elephant. At the same time, Sullivan entertained the idea of bringing mammoths back to life, hinting that while the possibility of cloning a mammoth was improbable, at least right now, it was not entirely impossible.[36]

A few years later, Wilson and colleagues were ready to publish, and in 1980 their paper in *Science* appeared. Ellen M. Prager, a molecular biologist and postdoctoral researcher with Wilson, and Alice Taylor, an electron microscopist, had found evidence of ancient proteins from Dima along with well-preserved microscopic muscle structure.[37] Reporter John Noble Wilford, also writing for the *New York Times,* described their research as an "exploratory tool in the emerging science of fossil genetics" but also speculated on its use as a potential tool for resurrecting the mammoth if only its DNA could be found: "If they could find intact strands of DNA . . ., the raw material of heredity, they could conceivably reconstruct the long-extinct species through cloning, though the chances of doing this are considered quite remote."[38] To be clear, Prager and colleagues did not find mammoth DNA, nor did they attempt to look for it in this specific study. It was not until five years later that Wilson and Higuchi attempted the task, and they only did so with difficulty. In the end, they were able to detect the presence of DNA from the baby mammoth but were unable to replicate or authenticate the DNA.[39] Nonetheless, early evidence for the preservation of molecules in ancient and extinct creatures, be it proteins or DNA, generated much speculation across the media about scientists' potential to bring extinct animals back to life.

It was not long before this far-out speculation about mammoth resurrection looked like a reality, at least according to the media. In April 1984, *MIT Technology Review* reported a story declaring that the mammoth, extinct for the past tens of thousands of years, had been brought back from the dead. A Dr. Yasmilov of the University of Irkutsk and a Dr. Creak from the Massachusetts Institute of Technology were the alleged masterminds behind this feat. According to the article, Yasmilov had recovered a frozen mammoth egg from a carcass in Siberia and sent the sample to Creak. Creak recovered DNA from the frozen mammoth egg, then combined the sequences from the extinct mammoth with sequences from the sperm of an extant Asian elephant. The resulting product was implanted into the wombs of several Indian elephants that served as surrogates for these elephant-mammoth hybrids. Although a number of the surrogates miscarried, two gave birth to the first elephant-mammoth hybrids. Scientists called them a new species, *Elephas pseudotherias.*[40]

The news went viral. The *Chicago Tribune,* for example, reprinted the report, which was subsequently sensationalized by hundreds of newspapers across the United States. In the end, however, the entire story was a hoax, written by an undergraduate student, Diane Ben-Aaron, for an undergraduate course. Although it was published on April 1, 1984, April Fool's Day, reporters had not noticed or made the connection. Regardless, the tale's instantaneous popularity across media outlets made it clear that the idea and perceived reality of bringing extinct species back to life was something the public wanted, or even needed, to know about. It spoke to the press's and public's simultaneous fascination with and fear of genetic engineering and technology, thus provoking a host of ethical, moral, political, and environmental arguments both for and against it.[41]

The potential to resurrect extinct species continued to hit headlines, with reports ranging from the wildly speculative to the more subdued. The weekly tabloid *National Examiner,* for example, published a report claiming, "Mad Scientists Are Cloning Dinosaurs as Weapons of the Future." The report contained a mostly false story combining mammoth cloning and American-Soviet nuclear warfare conspiracies with actual scientific research going on at Berkeley.[42] At the same time, there was more accurate reporting. The magazine *New Scientist,* for example, noted that "stories that the quagga, the dodo, and the mammoth might be about to rise and stalk the Earth once more are somewhat exaggerated," but "resurrecting the quagga" might one day "indeed be possible." In fact, the report's title—"The Resurrection of the Quagga"—hinted at as much.[43]

Importantly, media reporters were not the only ones talking about the resurrection of extinct species. Indeed, some scientists entertained the idea too. In another *New Scientist* article—"To Clone a Dinosaur"—Mike Benton, a paleontologist then at the University of Belfast and now at the University of Bristol, directly attended to the topic of resurrection, specifically dinosaur resurrection: "Will we ever be able to clone a dinosaur?" Benton's answer was not yes but neither was the answer no. Although it would not be easy to reliably recover dinosaur molecules, Benton suggested that very small amounts of protein are likely to remain intact in fossils as old as the dinosaurs that existed 65–245 million years ago. More likely than discovering ancient dinosaur proteins, however, was the prospect of bringing the quagga back to life. According to Benton, it might be possible to resurrect the extinct quagga by inserting its DNA into the embryo of a mountain zebra to create some sort of quagga-zebra hybrid.[44]

THE ROLE OF SPECULATION

Speculation was a central feature guiding various scientists' early research efforts to test the theoretical preservation and potential extraction of DNA from fossil species. Scientists from Poinar and Pääbo to Wilson and Higuchi clearly speculated about the recovery of DNA from fossil material hundreds to millions of years old. They wondered about its application to questions in evolutionary biology. They also wondered about its implications for changing how researchers study the past. Such conjectures were typical, even necessary, for generating and testing hypotheses, especially the most unconventional ones. In fact, philosophers of science have proposed that speculation is a useful and essential component of the scientific process. Adrian Currie and Kim Sterelny, for example, argue that speculation can be particularly productive in moving scientific inquiry forward. Although speculation by definition exceeds available evidence in support of it, "productive speculation" can be empirically grounded and hypothesis-generating.[45] It can increase interest and traction around an idea to ultimately produce convincing data in support of it. In these early research efforts to discover DNA in fossils, speculation motivated practitioners to undertake experiments that could generate evidence.

At the same time, there were more overt expressions of far-out speculation, namely around the hypothetical idea of species resurrection. Speculation extended beyond immediate research practices or potential. For example, media reporters, as well as some scientists, openly entertained the potential to use ancient DNA to not only study evolutionary history but to maybe one day bring back extinct species. To a certain degree, this type of speculation was productive in creating an awareness of and excitement for this new line of research for both public and professional audiences.

But speculation can only be so useful in promoting research. Moreover, speculation, or rather too much or the wrong kind of it, may even frustrate research efforts. Currie and Sterelny make this point, arguing that speculation, if idle, can be a vice. In these circumstances, idle speculation takes place when speculation cannot or does not advance research efforts by producing alternative scenarios or providing additional evidence needed to support those scenarios.[46] In fact, in these early years, scientists feared exactly this outcome, that too much speculation with too little evidence could harm their research and reputation in terms of credibility. Even Wilson, a visionary for his time, confessed that the search for DNA from fossils was an exotic and speculative undertaking. From the outset, Higuchi was concerned with contamination that would affect DNA authenticity and consequently its credibility. Yet even with reliable evidence, albeit preliminary evidence, for the successful extraction of DNA from the quagga, the NSF funding panel rejected Wilson and Higuchi's proposal on the grounds that the research and evidence for it was underdeveloped and not widely applicable across the sciences. Moreover, Pääbo conducted his study of ancient Egyptian mummies in secret for fear of failure and subsequent reprimanding or ridiculing. This tension between science and speculation—most notably scientists' awareness of and ability to engage or disengage as necessary—was an equally important element that colored ancient DNA's emergence and evolution into a new way to study evolutionary history.

This is not to underestimate the importance or impact of the quagga study, nor that of the mummy study, in marshaling wider attention and support for the search for ancient DNA. In the quagga and mummy studies, the evidence was impressive for the time, but within the bigger picture the evidence was also weak. A couple of short DNA strands from a single sample of an extinct quagga and one ancient mummy were exciting but not extra-ordinary. The preservation of DNA in some fossils did not guarantee the preservation of DNA in all fossils. Ancient DNA, at least at this time, was an anomaly. And the science behind it was more of a spectacle and a rarity than a predictable method of obtaining data. What was important about these studies was the acceptance of the evidence, however weak, by prestigious scientists and esteemed journals. Evidence, and professional publication of it, played a role in establishing a certain degree of credibility behind these novel ideas. However, in order for scientists to transform the search for DNA from fossils into a legitimate scientific research program, they would need better-preserved fossils, better techniques and technologies, more funding, more evidence, and lots of luck.

CHAPTER THREE

Testing Limits

THE POLYMERASE CHAIN REACTION

By the mid- to late 1980s, the search for DNA from ancient and extinct organisms was attracting both professional and popular attention, but scientists were aware of their need for better technology that could reliably amplify the decayed and damaged DNA characteristic of old specimens. They knew that new technologies and techniques were necessary if they wanted to transform the search for DNA from fossils into a full-fledged research program. Rather conveniently, the innovation of a new molecular biological technique, the polymerase chain reaction (PCR), coincided with this search for ancient DNA, presenting a convenient solution. PCR was first developed in the 1980s by Kary B. Mullis and fellow colleagues at Cetus Corporation, a biotechnology company in Berkeley, California. Following a presentation of its application at the Cold Spring Harbor Symposium in 1986—the same symposium where James Watson first described in detail the DNA double helix thirty years earlier—and several publications from 1985 to 1987, PCR became the most widely used technique in molecular biology.[1] As various researchers incorporated the technique into their work, it began to transform the field of molecular biology and the related areas of systematics, forensics, and medicine. In fact, PCR proved so revolutionary that Mullis was awarded the 1993 Nobel Prize in Chemistry for its invention.[2]

The advantage of PCR was its automatic amplification of DNA. This took the mental and physical strain out of the previously manual process of cloning using vectors to create multiple identical copies of the DNA of interest, as was the case when scientists first extracted DNA from the quagga. Overall, PCR could create billions of copies of DNA sequences from only a few strands, or even just one strand, of DNA. Ancient DNA was often preserved in short strands, and PCR was specifically well suited for amplifying these damaged and degraded fragments. Furthermore, PCR was quick and inexpensive. It used repeated cycles of heating and cooling to copy the DNA. First, heat would be applied to separate double-stranded DNA into single-stranded DNA. Then the single-stranded DNA would be exposed to primers. These primers would attach themselves to the appropriate sites of desired DNA to be amplified. Finally, a copy of the targeted DNA would be produced. This process would continue as a chain reaction, ultimately creating millions to billions of copies of the targeted DNA.[3] *New Scientist* called it "a tool of unbelievable power."[4]

As an early player in the search for ancient DNA, Allan Wilson's lab at the University of California, Berkeley, was the first to apply the technique to the study of old specimens. Wilson was already well connected with Cetus Corporation, where PCR had been designed and developed. Furthermore, Cetus was not too far from the university, so Wilson sent Higuchi to learn the method and bring it back to the lab.

Svante Pääbo—by this time a new postdoctoral researcher in the lab—set out to test PCR's utility on a range of specimens of different ages and from different environments.[5] Pääbo's goal in this study was twofold. First, he was interested in testing PCR's technical advantages. If PCR could be easily and reliably applied to old specimens, then this could have notable implications for using museum specimens and other archeological and paleontological samples to answer unresolved questions about evolutionary history. Second, he was interested in testing the limits of DNA preservation in various specimens and how the chemical composition of DNA could be affected by modifications that occur through the desiccation of tissues as a result of hydrolytic and oxidative processes. In better understanding the properties of DNA and processes that contribute to its degradation, he hoped to find observable and generalizable patterns.

With the availability of PCR, Pääbo extracted DNA from a four-year-old piece of pork, fragments of mummy material, the remains from an extinct wolf-like species called a thylacine, and a thirteen-thousand-year-old extinct ground sloth. To his surprise, he found that the age of the sample did not necessarily correlate to the amount of DNA preserved or the degree to which it was damaged. The preservation potential of DNA appeared quite fickle and definitely subject to internal and external processes of degradation. However, PCR greatly increased the possibility and ease of extracting and amplifying DNA sequences from ancient samples—but its very virtue, namely its ability to exponentially amplify even the smallest amounts of DNA, was also its vice. PCR was extremely sensitive in that it could detect and amplify a single molecule. This was convenient for ancient DNA because it was often preserved in short fragments. However, if there was any other DNA in the sample, especially DNA that had been introduced during the handling or studying of the specimen in the lab, then that DNA—being much more recent and better preserved—would be preferentially detected and amplified by PCR instead of the actual ancient DNA of interest. To control for contamination, Pääbo suggested "rigorous precautions" when preparing and handling samples, solutions, and materials in the lab.[6]

In a review article in the late 1980s, Pääbo, Wilson, and Russell Higuchi expanded on the implications of PCR as a technological development within the broader context of molecular evolutionary biology, especially highlighting the use of PCR in making the search for ancient DNA possible, as well as the risk of contamination that came along with it. According to them, the general frustration of molecular evolution was the current challenge of reconstructing evolutionary history with only the DNA from living organisms. Without DNA from ancient and extinct organisms, they claimed, it was difficult if not impossible to understand evolution over time. PCR, however, was one potential solution for overcoming this "'time trap.'" "The recently achieved ability to study DNA from museum specimens and archaeological finds via PCR," they wrote, "opens up the possibility of studying molecular evolution by actually going back in time and directly approaching DNA sequences that are ancestral to their present-day counterparts."[7] Sure enough, a growing body of evidence demonstrated PCR's utility in testing hypotheses in evolutionary biology.

Although PCR was a welcome innovation with many advantages, there were drawbacks to the technology that researchers had to first acknowledge in order to accommodate or avoid them. Pääbo, Higuchi, and Wilson took the lead on this, outlining a short but succinct list of criteria to control for

contamination when working with paleontological and archeological material. Here, they recommended three criteria. First, they suggested comparing
sequences obtained from an ancient organism to the sequences of its closest
living relatives, then using the accuracy of the resulting phylogenetic analysis as an indication of ancient DNA authenticity. Next, they advised using
control extracts to detect for contamination in any of the solutions or
reagents in the lab, as well as independent extracts in order to recover and
demonstrate authentic DNA sequences from more than one sample. Finally,
they recommended a "strong inverse correlation between amplification efficiency and size of the amplification product." In other words, they hypothesized that DNA from long-dead organisms should yield shorter sequences,
approximately 150–500 base pairs, rather than longer sequences. These
shorter sequences were expected to reflect the fragmented nature of ancient
material, while longer sequences might signal contamination from modern
material. They argued for the importance of understanding the potential for
contamination and the need to take steps to control for it in order to determine the authenticity and reliability of ancient DNA. "When the three above
criteria . . . are fulfilled," they argued, "a given sequence is considered likely
to be of ancient origin."[8]

With the newfound availability of PCR, Wilson, Pääbo, Richard H.
Thomas (another postdoctoral researcher in the lab), and Walter Schaffner
(a molecular biologist from the University of Zurich) next attempted the
extraction of DNA from the thylacine, an extinct wolf-like marsupial. The
last thylacine, also known as the Tasmanian tiger, died at the Beaumaris Zoo
of Australia in 1936. At the time of its death, no one seemed to care that this
creature was the last of the species *Thylacinus cynocephalus*. In fact, more
than five months passed before its death and the extinction of the species
were even noticed and announced.[9] The thylacine was an unusual animal—
wolf-like in face and body, marsupial-like in anatomy and physiology, carniv-
orous in appetite, and nocturnal in behavior. It had a kangaroo pouch and
resembled a tiger with a yellow-brown coat and dark stripes across its back.
Like the quagga, the thylacine was a sentimental species because of its
extinction, but it was also an obvious object of study for its mysterious evolutionary history. Indeed, systematists had long argued over its phylogenetic
placement. Some suggested a closer connection to an extinct group of South
American marsupials, while others considered the thylacine to be related to
Australian marsupials. The debate among scientists came down to the fossil

evidence and different interpretations of it. For example, the thylacine and South American borhyaenids shared similar dental and pelvic traits, while the thylacine and Australian dasyurids shared similar hind limbs.

In 1989, Thomas and colleagues published their findings on DNA from the thylacine. Initially, they had attempted to recover DNA from a number of samples but were only able to extract DNA from one of them. From this, however, they were able to sequence 219 base pairs of mitochondrial DNA. Despite being such a short DNA sequence, it provided enough material that the team could compare it to the mitochondrial DNA of six other marsupials. After analysis, they concluded that the thylacine was most related to Australian dasyurids, which include the Tasmanian devil. This suggested that the thylacine was native to Australia, not South America as some had supposed.[10]

Based on the thylacine DNA sequences, the question of its evolutionary history seemed resolved. For some scientists, however, the fossil data gave evidence of a different history. Morphological data, such as fossils, and molecular data, such as protein or DNA sequences, provide important but different kinds of information about an organism's life. While researchers consider both types of data when trying to reconstruct the evolutionary history of organisms, the information can be inconsistent. For the thylacine, for example, the genetic data suggested an Australian origin but the fossil data seemed more consistent with a South American origin. In this instance, the team sided with the genetic evidence and concluded that the thylacine, based on both DNA and protein evidence, originated in Australia.[11] Thomas and colleagues were confident in the fact that the sequences they recovered were authentic to the thylacine. To reconcile the inconsistencies between the fossil and genetic data, researchers explained the similarities between the Australian thylacine and South American marsupials as an example of convergent evolution, where two species evolve similar features independently of one another.[12] Overall, the thylacine study, like the quagga study, helped confirm the significance of studying DNA from ancient and extinct museum specimens, opening an unchartered territory of research.

FUNDING A NEW FIELD

Although the extraction, amplification, and sequencing of DNA from fossils were first explored in the United States, specifically at Berkeley, it was in the United Kingdom that ancient DNA research first received funding on a substantial scale. The quagga study provided evidence that DNA could be recovered from ancient specimens, but in doing so, it introduced more questions than answers, inspiring other scientists to join the search for DNA from damaged and degraded material.[13] In November 1988, the Natural Environment Research Council (NERC)—the largest funding body for the environmental sciences in the United Kingdom—awarded a £600,000 grant for the search for molecules in fossil material. Over the next four years, the "Special Topic in Biomolecular Palaeontology" would fund a host of scientists to search for ancient lipids, proteins, RNA, and DNA from a variety of specimens.[14]

"The Special Topic in Biomolecular Palaeontology" would prove to be an invaluable initiative in the conceptual, organizational, and financial development of ancient DNA research as a young scientific field. Indeed, it financed some of the earliest, perhaps most exploratory research of the time. Among the numerous applications the NERC received in response to this grant was an application from a young group of scientists proposing to extract DNA from ancient bone—and not just any bone but ancient human bone. One interviewee involved in the review process recalled this particular proposal, as well as their immediate reaction to it: "I had a look at it, and it was *the most stupid idea*. It was this young team and they wanted to get DNA from fossil bones." The proposal, at least according to this scientist, was absurd. One reason had to do with the available evidence for the longevity of not just DNA but other molecules such as proteins. At the time, proteins were not expected to survive intact for thousands, let alone millions, of years. "DNA is much less stable than proteins," explained this interviewee. "There's no way you could get DNA to survive in fossil bones" (Interviewee 9).

There were a number of other reasons this proposal seemed so far-fetched. Although it might be more likely to recover DNA from skin or muscle tissue, as was the case with the quagga and mummy studies, it seemed less likely scientists could recover DNA from a substance like bone. A fully mineralized fossil, such as an ancient human bone, would be highly unlikely to preserve DNA, because its organic components would have decayed and been replaced by minerals from the surrounding sediment in which the organism died. Furthermore, even if DNA could survive the test of time and remain untouched in a substance like bone, it would be extremely difficult to verify the DNA's authenticity. That is, it would be next to impossible to demonstrate that the ancient human DNA extracted from the ancient human bone was not a contaminate from being handled by human curators and researchers in the lab.

As the reviewers continued assessing the proposal, however, they came across some important information that made them change their minds. Indeed, the applicants had some rather convincing empirical evidence to back up their proposal. According to a researcher, just as they were disputing the proposal's feasibility, a fellow colleague and reviewer excitedly pointed to one of the pages: "Look! . . . We've actually got a gel." Surprised, this interviewee took a look and quickly replied, "Oh! Well, if they got the band from the gel we should give them funding!" (Interviewee 9). What the reviewers found themselves looking at was photographic evidence of DNA, a run of small dark bands spread across a gel-like substance. It was a classic experimental technique—gel electrophoresis—for the purpose of visualizing DNA fragments, should they in fact be present. The researchers had inserted a dye into the extracted DNA sample, then inserted that sample into a small rectangular gel. Afterward, an electric current reacted through the gel, moving the DNA along the gel and separating the shorter fragments from the longer ones. For the reviewers, this was evidence enough. In the end, "We gave them a positive review," explained this interviewee, "and that funded Erika Hagelberg. And the band [on] the gel was the band [on] the gel that then appeared in *Nature* as the first record for DNA recovery from old bones" (Interviewee 9).

The next year, 1989, *Nature* published a paper by Erika Hagelberg, Bryan Sykes, and Robert Hedges at the University of Oxford on the "successful extraction and amplification of DNA from human bones between 300 and 5,500 years of age."[15] Reminiscing on the impact of this research at the time, another interviewee and pioneer in this research referred to the findings and their subsequent publication as a "watershed" moment. It provided evidence that some DNA could, and in this case did, survive in paleontological and archeological material including bones, not just skins and tissues. "Twenty-five years ago," the interviewee said, "people had no idea whether DNA survived in bone, and if it did what to do with it or how to get it out in the first place" (Interviewee 11). This study, the first of its kind, elicited excitement. It generated a good deal of skepticism too.

Despite publication in a top-tier journal such as *Nature,* some scientists found it hard to believe the results. In 1990, the controversy over the preservation and extraction of DNA from old bones came to a head when Pääbo directly confronted Hagelberg at the Biomolecular Palaeontology Community Meeting at the University of Glasgow.[16] "Svante Pääbo, very famously at the meeting, stood up and said, 'Of course, you can't get DNA from bone!'" recalled a researcher. And that was "just before Erika Hagelberg stood up and said, 'Here's my results on DNA from bone'" (Interviewee 9). Another scientist recounted a similar situation: "Svante had . . . some very public fights with her in conferences . . . saying it was all shit." According to this scientist, "Svante stood up and said, 'This is shit! It's full of shit! Where are your controls? You haven't got any! And the sequences you have are rubbish!'" In the end, "it was a big shouting match" (Interviewee 32), and "she [Hagelberg] felt very much as if he [Pääbo] was trying to undermine her work at the time" (Interviewee 9).

The fight over DNA from old bones came down to concerns about contamination. With human scientists working on human remains, it would be difficult, if not seemingly impossible, to detect modern contamination via the traditional sequence comparison. In fact, this controversy over contamination, specifically as it related to the study of ancient humans, would impact the young field for years to come. Importantly, it was not just the controversy itself at the time but the retelling of it to fellow colleagues and later students that served to establish and reinforce a specific narrative of the early days of the field. "For a long time," said a younger practitioner in the field, "ancient DNA [research] was about, 'What is possible?' 'Is it possible—in the very early days—to get DNA from bone?'" Although not present at the conference where Pääbo confronted Hagelberg, this scientist did recall the retelling of it: "I wasn't there at the time," they said, "but I heard the story that at one of the first ancient DNA meetings Svante Pääbo said you will never be able to get ancient DNA from bone" (Interviewee 15). These disagreements were far from superficial. As researchers continued to test the limits of this new field, the issue of contamination would continue to define, even drive, the development of the field.

Sure enough, some researchers tested the limits to see just how far back in time DNA could survive. In 1990, for example, Edward Golenberg at the University of California, Riverside, and fellow colleagues reported the recovery of the oldest DNA to date: 17–20-million-year-old DNA from a fossil *Magnolia* leaf recovered from the Clarkia deposit in northern Idaho.[17] The research findings, published in *Nature*, marked the recovery of the oldest DNA to date. An article in the *Washington Post* noted, "Scientists for the first time have read the genetic code of an organism that died between 17 million to 20 million years ago, achieving a record-breaking glimpse into the past based on new techniques that could soon be used on other ancient plants and animals."[18] The *New York Times* ran a report, "Genetic Code Found in 17-Million-Year-Old Leaf," quoting a scientist who called it a "fantastic breakthrough" and the NSF, which claimed it was an "unprecedented achievement."[19] *New Scientist* devoted an entire six pages to the discovery of "The Oldest DNA in the World," which "has left molecular palaeontologists with more questions than answers."[20] Indeed, some researchers had serious questions about the authenticity of the multimillion-year-old DNA.

Pääbo and Wilson, for example, were wary of the results, noting that the recovery of such DNA "seems to surpass our wildest dreams."[21] They were suspicious of Golenberg and colleagues' results mainly because the sequences he claimed to have recovered were too long. At 790 base pairs, this sequence surpassed Pääbo's suggested threshold of 150–500 base pairs, which would be more characteristic of old DNA. In a separate study, Pääbo, Wilson, and Arend Sidow—a colleague from the University of Munich—set out to replicate the results. They did, in fact, recover evidence of DNA. However, the DNA was not plant DNA but bacterial in origin. Unable to replicate Golenberg's results, and in light of the fact that they determined their own extractions to be bacterial in origin, Pääbo and colleagues suggested that more extensive work be conducted to demonstrate the veracity of claims around multimillion-year-old DNA.[22]

In 1991, Pääbo and Golenberg came face to face at the Biomolecular Palaeontology Discussion Meeting at the Royal Society in London, both giving presentations on their separate findings of the Clarkia deposit.[23] Martin Jones, an archeologist at Cambridge University and another early researcher in the search for DNA from fossils, documented the tension in the room as the two young researchers presented their conflicting conclusions. As Martin recalled, Golenberg went first and in a "slightly nervous presentation" discussed the need for "care and control" but stood by his results of having obtained ancient and authentic DNA from a fossil leaf. Pääbo went next, explaining that in his own work on samples taken from the same site they had not recovered plant DNA but bacterial DNA. According to Martin, "The inference was that Golenberg's result arose from contamination."[24] Indeed, contamination was an ever present concern: one step forward was almost always accompanied by two steps back.

The growing funding and awareness of ancient DNA research, as well as the availability of PCR, invited more scientists to join the search for ancient DNA. This increased competition brought about bolder claims of long-term DNA preservation in a variety of material, high-impact scientific publications, and high-profile media attention. It also brought about more accusations of contamination. "Up until the *Magnolia* publication," wrote Martin, "the front runner in the race for ancient DNA was emerging as Svante Pääbo." However, as other scientists joined the hunt, often testing the limits as they did so, Pääbo found his role in the field shifting. According to Martin, "He was no longer simply the bright young star of the field, but was getting used to a new role as traffic policeman in a convoy moving with rather too much momentum for its own safety."[25] Ancient DNA research was evolving into a new field of scientific inquiry but one colored by controversy. For the skeptics, extraordinary claims required extraordinary evidence.

SCIENCE AND FICTION COLLIDE

In July 1991, the University of Nottingham in England hosted a conference called "Ancient DNA: The Recovery and Analysis of DNA Sequences from Archaeological Material and Museum Specimens."[26] It was the first official international meeting of its kind. Richard Thomas—formerly at the University of California, Berkeley—had recently relocated to London as director of the DNA Laboratory at the British Museum of Natural History and was responsible for organizing the event. The conference's objective was to bring together an international and multidisciplinary group of practitioners interested in ancient DNA research to share results, compare research, and discuss potential protocols as well as problems.

The two-day conference—sponsored by the British Museum of Natural History along with NERC and Cetus Corporation—featured nearly thirty-five presentations by researchers from the United States, United Kingdom, Netherlands, Germany, Italy, Israel, Denmark, Sweden, France, Spain, and South Africa. Overall, the meeting involved senior and junior researchers alike from different disciplines ranging from archeology, paleontology, and geology to molecular biology, genetics, and forensic science.[27] Some of the researchers demonstrated the use of ancient DNA to trace the evolution and domestication of plants, while others tested hypotheses about the evolutionary relationships of extinct and endangered animals. Others focused on human evolutionary history, namely the sexing of skeletons for kinship.

According to one interviewee, an early leader in this research area, "Everyone was really excited. It was a completely unmapped field—getting DNA from dead things. No one had ever done it before." Still, the enthusiasm for the new field was not entirely unbridled. While there was "a lot of really ambitious speculation," there was also "a lot of realism about what could be done and what couldn't be done" (Interviewee 4). Indeed, researchers recognized that not every specimen was going to yield DNA, and if it did, it would be degraded or damaged, and difficult to determine its authenticity.

The public had quite a different view. In November 1990, nearly six months before this meeting, Michael Crichton's science-fiction novel *Jurassic Park* was published.[28] The book—whose plot was based on scientists' recovery of dinosaur DNA from amber-preserved insects—was an immediate success, and after being translated into multiple languages from Chinese and Japanese to Hungarian, it became an international best-seller. There was also much excitement and anticipation around turning the book into a Hollywood movie. Crichton was an already famous author whose novels had been made into highly successful movies. According to Don Shay and Jody Duncan, authors of *The Making of* Jurassic Park, Crichton sent his manuscript to his publisher, Alfred A. Knopf, in May 1990. In no time at all, major movie producers from Twentieth Century Fox and Warner Brothers to Universal Pictures were jockeying for the chance to make the film. However, as Shay and Duncan noted, Crichton had privately promised Steven Spielberg the film rights, an agreement that reportedly transpired when Crichton and Spielberg worked together on a screenplay, *ER*, which Crichton had written and Spielberg was at the time crafting into a medical drama television series. Nonetheless, intense bidding ensued. In the end, the film rights went to Universal and the job of producing and directing went to Spielberg.[29] With such an award-winning director in charge, *Jurassic Park* was a highly anticipated blockbuster.

The enormous popularity of *Jurassic Park* was partly a result of its scientific and technological plausibility.[30] After mulling over the *Jurassic Park* story line for nearly a decade, Crichton settled on a science-fiction thriller about genetically engineered dinosaurs brought back to life from DNA preserved in the gut of a mosquito trapped in ancient amber. Jurassic Park was a world-class dinosaur theme park as well as a cutting-edge scientific experiment gone wrong. In his book, Crichton utilized current DNA technology and research on cellular and molecular preservation in fossils to make the story convincing.

Specifically, Crichton used the work of George Poinar and Roberta Hess at Berkeley as the novel's premise. Poinar and Hess, advocates for the search for molecules from fossils since the early 1980s, had since teamed up with Allan Wilson and Russell Higuchi to try to recover DNA from amber insects.[31] Although Poinar and Hess's research was influential, it was the quagga research by Wilson and Higuchi, as well as the early mummy study by Pääbo, that made *Jurassic Park* so plausible to the public. "Genetic material had already been extracted from Egyptian mummies, and from the hide of a quagga, a zebra-like African animal that had become extinct in the 1880s," explained Crichton in his book. "By 1985, it seemed possible that quagga DNA might be reconstituted, and a new animal grown. If so, it would be the first creature brought back from extinction solely by reconstruction of its DNA. If that was possible, what else was possible? The mastodon? The saber-toothed tiger? The dodo? Or even a dinosaur?"[32] This research, and the speculation about bringing extinct creatures back to life associated with it, was the backdrop for International Genetics Incorporated, the fictional bioengineering company in Crichton's story, and their remarkably successful efforts to clone full-size dinosaurs.

In June 1991, just before the Nottingham conference that July, the science section of the *New York Times* made note of the upcoming meeting, advertising it alongside a recipe for bringing dinosaurs back to life. The report—"Scientists Study Ancient DNA for Glimpses of Past Worlds"—was written by the well-known science reporter Malcolm W. Browne. "Will it one day become possible to breed a living dinosaur from genes preserved in fossils?" he asked. "Although most scientists regard such an idea as unrealistic, a few have begun to conclude that it can no longer be dismissed out of hand." In a step-by-step illustration, Browne outlined a "Recipe for a Dinosaur." The first step: "Find a bead of amber that contains a blood-sucking insect from the age of the dinosaurs." Next, "extract genetic material from blood cells of a bitten dinosaur, and amplify DNA with the PCR technique." Then, "process and inject into embryo of an alligator." Last but not least, "wait until it hatches." Browne credited this "recipe for a dinosaur" to George Poinar, citing it as the inspiration for "the basis of the best-selling science-fiction novel 'Jurassic Park.'" "Obviously, we couldn't reconstruct an extinct animal today, even if we had all its DNA," Poinar said in an interview with Browne. "However, my belief is that there are dinosaur cells inside biting flies trapped in amber of Cretaceous age and older. It's just a matter of finding the dinosaur DNA and getting it out."[33]

Although the first official and international ancient DNA conference put the practice on the map for a professional audience, *Jurassic Park* and the *New York Times* story put the science in the media spotlight. In the "Research News Series" for *Science*, Jeremy Cherfas covered the conference, highlighting the fact that the meeting attracted more attention than anticipated. According to Cherfas, Thomas, one of the conference organizers, had hoped for a "quiet" and "technical" meeting, "but that was before the science section of *The New York Times* published a fanciful 'recipe' for recreating a dinosaur from ancient DNA." "We were inundated by people," Cherfas reported Thomas saying. "We were stunned and amazed by the reaction from the press. We had to spend a fair amount of our time telling them, 'No, we are not going to reconstruct the dinosaur.'" "However much scientists may protest that it cannot be done," Cherfas wrote, "the public and the popular press clearly expect ancient DNA to create Jurassic Park for real."[34]

FOUNDING A NEW FIELD

By 1991, a scientific community had come together under the name of "ancient DNA research" and started to communicate professionally about their expectations for what appeared to be a new research field. Reporting for *Science,* Cherfas interviewed Thomas, who noted that earlier meetings on ancient DNA research had been "controversial" as "scientists disputed the validity of their techniques." As far as Thomas was concerned, the University of Nottingham conference was different: "There, he [Thomas] says, with considerable satisfaction, people at long last talked openly about the problems they had with their samples." Indeed, the community appeared to be moving toward more open, honest, and collegial discussions of the potential, as well as pitfalls, of the search for DNA from fossil specimens. According to Cherfas, for the conference attendants, the take-home message of the meeting was clear: "They found they had created a new field."[35]

In addition to the conference, the ancient DNA community was interested in taking other measures to establish and expand on the work being done. At this conference in particular, several attendants considered the idea of forming a research journal dedicated to the search for DNA from fossils.

Ultimately, they decided against it for the time being, reasoning the field might be too premature to provide sufficient content and ensure long-term success. Instead, they settled for a newsletter to better connect researchers both professionally and personally.[36] Robert Wayne, an evolutionary biologist at the Zoological Society of London, and Alan Cooper, a graduate student at the University of Wellington in New Zealand who had just started working at Berkeley with Wilson and Pääbo, accepted the "dubious honor" of becoming the newsletter's first editors.[37]

In a letter dated April 1992, Wayne and Cooper introduced the first *Ancient DNA Newsletter*. The newsletter included up-to-date summaries of research projects, study outcomes, and practical lab tips. There was even a question-and-answer section—"Dr Russ' Problem Corner"—where scientists could share their technical lab troubles with Higuchi, the molecular biologist guru, and receive a response in the following issue. At the same time, the newsletter was more than just business. The "Personals" of each newsletter included "general gossip," from write-ups of special events and restaurant reviews to short research statements with the "intent of building bridges between laboratories with common interests."[38] To help fund the newsletter's continued production and aid the success of the new field, researchers could buy their very own "Ancient DNA" shirt featuring the *Ancient DNA Newsletter* logo, a *Tyrannosaurus rex* holding a pipette in one hand and a double helix in the other.[39]

Specifically, this newsletter was a space for scientists to construct a culture of professional and philosophical values around the practice. This was especially important because the search for ancient DNA brought together scientists from disparate disciplines, from archeology and anthropology to botany, paleontology, molecular biology, and forensic science. Researchers had diverse motivations for joining the search, as well as their own unique sets of research methods, questions, and traditions. "It was a way of really standardizing the techniques and information and methods that were going on in the field," recalled one researcher. "And it was pretty important at the time in terms of solidifying the field as an entity rather than people just using ancient DNA for quite different things" (Interviewee 32). In light of their disciplinary differences, this early community of ancient DNA researchers had a common interest in reliably extracting, sequencing, and analyzing DNA from long-dead organisms. Ancient DNA and the pursuit of it was their common ground, their boundary object, and the newsletter was a means that

Cover of *Ancient DNA Newsletter* (April 1992),
illustrated by the biologist and paleontologist Blaire
Van Valkenburgh. (Courtesy of Blaire Van Valkenburgh)

scientists used to consolidate their various educations, experiences, and train-
ings toward the pursuit of this shared goal.

The first ancient DNA conference and subsequent newsletter, as well as
earlier funding efforts and high-profile scientific publications, were fundamental to the emergence of ancient DNA research as a new field. At the
same time, the media played a real role in its formation too. The media
helped bring attention to a marginal, rather speculative idea, and in doing so
invited both professional and popular interest in its potential as a revolutionary approach to studying evolutionary history. Specifically, Crichton's
best-selling book and movie in the making corresponded with some of the
field's initial funding efforts, publications, and conferences. The *New York
Times*'s explicit connection between the first ancient DNA conference and
Crichton's *Jurassic Park* positioned the new field and its practitioners in the
media spotlight. For the public, *Jurassic Park* gave life to the rather abstract
concept of extracting DNA from long-dead creatures.

For the majority of scientists, however, the goal was not to bring dinosaurs, or any other extinct creature, back to life. Higuchi and Pääbo, for
example, discussed species resurrection with media reporters only to reinforce the point that it would be impossible, impractical, and even unethical.

In his *Science* report, Cherfas speculated about resurrecting a woolly mammoth based on the fact that Wilson's lab had earlier recovered some small sequences of DNA. Cherfas asked Higuchi his thoughts on the topic: "The amount of mammoth DNA is enough that in theory a dedicated graduate student could reassemble the entire mitochondrial genome," Higuchi said. "So we could have elephants walking around carrying mammoth mitochondrial DNA sequences." According to him, however, there was problem with this: "It would make absolutely no difference. They'd still be elephants."[40]

Pääbo also argued that such a project would be far from worthwhile. In a different media article, he told reporter Malcolm Browne, "It's theoretically possible to isolate the gene for a certain character, and introduce it into another species, if you thought that was worthwhile, which I do not." As far as he was concerned, he objected to species resurrection for practical and philosophical reasons: "You could find the gene for the typical quagga color pattern, for example, and introduce it into a zebra." However, "you would end up with something that looked like a quagga, but in reality it would just be a zebra that looked like a quagga."[41] Regardless of whether scientists were actually attempting to bring dinosaurs back to life, and despite their efforts to brush the prospect aside, the idea of resurrection and its connection to the search for DNA from fossils, especially as embodied by *Jurassic Park,* generated attention and activity around what was an otherwise novel research practice.

Even as scientists rejected the conclusions or implications of the *Jurassic Park* narrative, as well as media reporters' mentions of it, they drew on its popularity to bolster their own work. "The 'clone-me-a-dinosaur' faction," recalled one researcher, "people are obviously going to get excited about that . . ., and there were a few of us, I think, at the time that were happy to piggyback off that interest to get funding and so on" (Interviewee 24). Indeed, researchers were aware of the news values around their practice and some catered to it for the pragmatic purpose of creating awareness, which would hopefully bring more resources such as publications, collaborations, funding, and status. The connection between the science of ancient DNA research and the science fiction of *Jurassic Park* made this easy to do. Indeed, this connection between the two would become increasingly important as the field continued to evolve over the next few years.

Overall, the conference and the momentum it engendered seemed to be a success. It was an eventful and exciting time for the growing community, but also a somber time with the early and sudden passing of one of its founding members. Wilson, a pioneer in molecular evolution and an early player in this research, was diagnosed with leukemia and had been undergoing treatment. "Allan's lab . . . , it's the birthplace of ancient DNA," said a former student and colleague. "That lab was incredibly creative. You could come up with any hare-brained idea. Allan . . . encouraged it actually. . . . The crazier you were the more he encouraged you" (Interviewee 18). Wilson, unable to attend the conference, was not forgotten. According to another early researcher, the conference attendants took the time to make him a get-well card that they all signed (Interviewee 30). He died just two weeks later.

THE ROLE OF SPECTACLE

By the early 1990s, a small but growing group of researchers became interested in testing the limits regarding the theoretical preservation and potential extraction of DNA from museum specimens, human remains, and fossil material. Crucially, scientists' pursuit of this goal was chiefly facilitated by the innovation of PCR. With PCR, they were able to more easily extract and sequence the small amounts of degraded DNA from ancient and extinct organisms. As PCR's utility became apparent, they explored a variety of samples from ancient muscle and skin to even partially fossilized bone. Researchers were also interested in understanding more about what types of environments might yield DNA at both a higher quantity and quality. They investigated the likelihood of recovering DNA from specimens thousands to millions of years old. Meanwhile, some scientists sought to use ancient DNA sequences to investigate the properties of and processes that contribute to DNA degradation in hopes of finding observable or generalizable patterns. In all of this, the early community of ancient DNA researchers were testing the limits of DNA preservation and of a new field they had recently created.

The first ancient DNA conference at the University of Nottingham and subsequent founding of the community newsletter were landmark events that both aided in and signified the establishment of a new field. Additionally, the fact that some of the earliest experiments were funded by prestigious agencies and initiatives such as the United Kingdom's NERC Special Topic in Biomolecular Paleontology, and were published in top-tier journals from *Nature* to *Science*, demonstrated to the broader scientific community, as well as the public, that ancient DNA research was a promising endeavor in its own right. Indeed, these activities, achieved through the actions of the individuals who initiated them, denoted a move toward disciplinary development. As historians, sociologists, and other scholars have shown, the establishment of conferences, journals, newsletters, funding opportunities, employment opportunities, student training, and publications are all classic indicators of discipline formation.[42]

However, as the search for DNA from fossils developed into a discipline, it did so under the influence of intense press and public interest, particularly as this new line of research coincided with the publication of *Jurassic Park*, accelerating the attention that the field and its practitioners received. In testing the theoretical limits of DNA preservation, along with the technological limits of PCR, the early community of ancient DNA researchers showcased the discipline's potential as a way to study evolutionary history. As they did so, ancient DNA researchers realized the opportunity to harness the media attention that often accompanied the search for DNA from fossils, using it to their professional and personal advantage. In fact, doing so seemed expedient, even necessary. The discipline was a young one, and scientists had yet to prove its worth. It seemed that any publicity was good publicity.

For practitioners, opportunities for media attention were easy to come by. The recovery and reporting of DNA from fossils became a sort of spectacle that captured the imaginations of both professional and public audiences. While most spectacles are visual or audible phenomena, science can, and often does, take on other forms of show.[43] Ancient DNA, for example, was far from a sensory phenomenon. Rather, the act of procuring DNA from ancient and extinct species was a curiosity because doing so defied expectations of how long DNA could last and how it might be used in research. For some, ancient DNA was a way to directly study the past. For others, it was a way to bring extinct species back to life. Indeed, the very prospect of seeing a resurrected dinosaur, mammoth, or other creature was a grandiose spectacle.

Although practitioners had made strides toward the founding of a field, the recovery of DNA from fossils was still more of an anomaly than a predictable research outcome. Indeed, the preservation of DNA in some fossils did not guarantee the preservation of DNA in all fossils. At this point, researchers also realized that even if they were able to consistently extract DNA from old material, they would have to address the problem of contamination. PCR was extremely sensitive, with a tendency toward amplifying contaminating

DNA sequences as opposed to the ancient DNA sequences of interest. As early as 1989, Pääbo, Higuchi, and Wilson had suggested a short list of steps to take in the lab to circumvent contamination and to project a level of realism about the field's potential among its challenges.

In light of this, some scientists felt the need to counter, or at least control, the growing enthusiasm for ancient DNA research. It was important to scientists to maintain some sort of power over the direction of the discipline because a lot was at stake, namely their credibility. Consequently, practitioners found various ways to balance the science of ancient DNA research with the spectacle and speculation that seemed to follow it. They very much realized the importance of doing so in a practice that from its beginning held such a strong public appeal. Interestingly, the spectacle of extracting ancient sequences from some of the world's oldest organisms, and the press and public attention associated with it, would simultaneously help and hinder the growth of ancient DNA research as scientists tried to transform it from a curious phenomenon into a credible practice. As the discipline developed, and as scientists continued to test its limits, a handful of practitioners would make more deliberate and influential attempts to build boundaries around the practice, especially in response to contamination concerns and increasing media coverage.

CHAPTER FOUR

Dinosaur DNA

Following the publication of Michael Crichton's *Jurassic Park* in 1990 and the first ancient DNA conference in 1991, a handful of scientists tried their luck at the recovery of DNA from insects in ancient amber. With the added advantage of PCR, David Grimaldi, an entomologist at the American Museum of Natural History (AMNH) in New York, teamed up with Rob DeSalle, a molecular biologist at the same institution, to test the idea. Not only was DeSalle an expert molecular biologist but he had also worked as a postdoctoral researcher with Allan Wilson nearly ten years earlier. Indeed, he was familiar with some of the earliest amber experiments performed at Berkeley on this very topic. According to one researcher's memory of the collaboration, *Jurassic Park* was the impetus behind the study in the first place: "All I know is that [a colleague] walked into my office one day and said, 'Have you read *Jurassic Park*? We should try it. Let's crack open some insects and see if we can get DNA out of it'" (Interviewee 17).

Grimaldi and DeSalle's study, however, was more than an attempt to test the *Jurassic Park* hypothesis. Researchers were interested in evolutionary questions. As with the quagga and the thylacine, the evolutionary history of mastotermes, a genus of termite, had puzzled researchers concerning their relatedness to other insects. "We wanted something of phylogenetic significance," explained one of the researchers on the project, "not just [asking] 'Is there DNA?'" (Interviewee 17). There was a clear biological question about this organism's evolution and extinction that accompanied the technical task of extracting, sequencing, and amplifying the DNA.

Early in 1992, Grimaldi and DeSalle—along with colleagues John Gatesy and Ward Wheeler at the AMNH—extracted and sequenced DNA from an approximately 30-million-year-old termite, *Mastotermes electrodominicus*. Contamination was a known issue, and the team took prescribed precautions including negative controls, extraction blanks, and phylogenetic comparison to affirm DNA authenticity. Confident in the reliability of their results, they wrote up their research for publication. That autumn, *Science* published their paper as evidence of "the oldest DNA extracted from a fossil."[1]

Not surprisingly, their research received widespread media coverage.[2] According to one practitioner, there was "an enormous amount of media requests" for "writing" and "filming" on the topic of "ancient DNA." The AMNH and its practitioners benefited from the attention. "The AMNH had built our first molecular lab. 'Wow! See what our molecular lab has just done!' It was maybe a few years old and these results were coming out of the molecular lab," explained an interviewee. "So, the museum got a lot mileage out of it." The AMNH also optimized the opportunity for publicity by commissioning a traveling exhibition titled "Amber: Window to the Past."[3] "There is *no question* that it rode on the heels of *Jurassic Park*," said this same interviewee. "The museum played it up. Everyone did. There was a lot of promotion" (Interviewee 17). Individuals took part in the publicity too, such as Grimaldi publishing a book on the topic and writing a feature for *Scientific American*.[4] Both scientists and scientific institutions alike were quick to capitalize on opportunities for publicity.

From New York to California, the search for DNA from insects in ancient amber continued. This time Poinar teamed up with his son, Hendrik N. Poinar, a student at California Polytechnic State University, and Raúl J. Cano, a microbial ecologist at the same university. As far as one interviewee was concerned, the collaboration was "serendipitous." "*Jurassic Park*, the book, had just come out," recalled this researcher. "And [a colleague] asked, 'Would you be willing to help me test the *Jurassic Park* concept and extract DNA from amber?'" "I'm not really one to back out from a challenge," explained this interviewee, "so I said, 'Sure. Let's do it'" (Interviewee 31).

The study was difficult but rewarding. "I think the first challenge was really getting the DNA out and without presumed environmental contamination," said one of the scientists. "And the second and most important thing was trying to convince ourselves that what we were getting was actually real DNA." In the end, Cano, George Poinar, and Hendrik Poinar were able to demonstrate evidence for the preservation and extraction of DNA from a 25–40-million-year-old bee, *Apidae: Hymenoptera*, preserved in amber. "The most exciting thing was seeing the first faint band in the gel after a PCR" (Interviewee 31). With these seemingly positive results, they wrote up their conclusions for publication in *Medical Science Research*.[5] This study's results at least equaled, if not surpassed, Grimaldi and DeSalle's previous publication on the oldest DNA to date.

The media framed these near back-to-back publications in terms of a rivalry between two teams in what appeared to be a race for the oldest DNA. In a *Science* media news article, "30-Million-Year-Old DNA Boosts an Emerging Field," Virginia Morell wrote, "In their effort to be the first to extract and amplify DNA from amber specimens, Grimaldi and his colleagues have arrived at the finish line in a virtual dead heat with George Poinar."[6] Boyce Rensberger wrote in the *Washington Post*, "Rival research teams have found that fossil insects embedded in amber for as long as 30 million years still contained DNA fragments. . . . The DNA samples, found in extinct species of termites and bees, are said to be the oldest yet discovered."[7] Along with establishing the search for multimillion-year-old DNA in terms of a race and rivalry, these media reports also made a clear connection between the emerging science of ancient DNA analysis and the science fiction of *Jurassic Park*.

Indeed, the close connection between both of these studies and the best-selling book and upcoming movie was far from lost on the media. The *Washington Post* issued the headline "Entombed in Amber: Ancient DNA Hints of 'Jurassic Park'" and referred to both recent studies as "a case of science imitating art imitating science."[8] Even scientists recognized the interplay between science and science fiction in their own work. "We're a long way from recreating a termite," said DeSalle in an interview with Morell for *Science*. "At this stage we've extracted only a fraction of a gene." As far as DeSalle was concerned, "It's obviously science fiction."[9] Others, however, were not as quick to dismiss the possibility of resurrecting ancient and extinct species. Writing for the *New York Times*, Malcolm Browne raised the possibility of extracting DNA not just from insects in amber but from dinosaurs themselves: "Paleobiologists and science fiction buffs dream of obtaining DNA still older than that recovered from ancient termites and bees—perhaps even DNA from dinosaurs." In this article, Browne quoted George Poinar, who was more than willing to entertain such an idea: "'Sooner or later,' Dr. Poinar said, 'we're going to find amber containing some biting insect that filled its stomach with blood from a dinosaur before getting trapped in the resin that eventually turned into amber. The blood may contain actual dinosaur DNA. That will be an exciting discovery.'"[10]

JURASSIC PARK ON THE MOVIE SCREEN

In 1993, Cano and colleagues, along with George and Hendrik Poinar, set out to beat their own record for the world's most ancient DNA from amber insects. Their specimen of study was a 135-million-year-old amber-encased weevil, *Nemonychidae coleoptera*, dating back to the Mesozoic Era, when dinosaurs roamed the earth. After cracking open the amber fossil and taking some tissue from the insect's body, the scientists performed DNA extraction and amplification via PCR, then sequenced two short strands of DNA, one with 315 and the other with 226 base pairs. They used phylogenetic comparison with sequences from five different but closely related species to determine the authenticity of the DNA. In the end, they felt confident they had recovered authentic DNA from this ancient insect, and that they had successfully sequenced the oldest DNA to date, a point they made clear in the article they submitted to *Nature*. On June 10, 1993, *Nature* published their paper—exactly one day after the *Jurassic Park* movie premiere and one day before its public release in theaters across the United States.[11]

The media certainly recognized the timing between the scientific publication and the movie release. Browne, who often covered ancient DNA research for the *New York Times*, commented, "The report of the achievement is being published today in the British journal *Nature*, one day before the opening of 'Jurassic Park,' a much-publicized movie based on the notion of cloning extinct dinosaurs from their surviving DNA."[12] The timing of events resulted in widespread publicity. In a memoir, George Poinar and Roberta Hess Poinar also noted the media attention this paper produced, recalling how their research became instantly popular and that their findings were reported in over two hundred newspapers in the United States and four hundred newspapers globally. However, they also claimed this was all mere "coincidence."[13] Coincidence or not, the news hit the headlines and left the impression of intentionality.

A number of scientists, internal and external to the field of ancient DNA research, remarked on the timing too. Indeed, some saw it in a far from positive light. One practitioner and competitor in the field, for example, observed, "I thought it absolutely extraordinary that a scientific journal—there was no way it was a coincidence—that *a prestigious scientific journal* like *Nature* would hold on to an article to wait for the opening day of a movie. . . . Of course, that caused a *huge* media splash" (Interviewee 17). Given that *Nature* was, and is, a popular and commercial journal in addition to a scientific one, it was not entirely unreasonable, or unexpected, that the journal would take advantage of such an opportunity. However, this interviewee's perception that the timing was surprising at best, and unwarranted at worst, reflected the feeling that some scientists viewed popular outward influences on science as potentially damaging to the integrity of the science.

At the same time, others viewed this interplay between science and media as a positive phenomenon. The late paleontologist and science writer Stephen Jay Gould, for example, offered his own observations about the timeliness of it all: "The nearly complete blurring of pop and professional domains represents one of the most interesting spinoffs—a basically positive one in my view—of the *Jurassic Park* phenomenon. . . . When a staid and distinguished British journal uses the premiere of an American blockbuster to set the sequencing of its own articles, then we have reached an ultimate integration."[14] Regardless of the positive, negative, or questionable consequences, press and public attention was a crucial component in the growth of ancient DNA research in terms of raising awareness of this novel and niche but increasingly high-profile activity.

If the book was an instant success, then the cinema production of it was a massive triumph. Even before the movie release, *Jurassic Park* was backed by a colossal marketing campaign. Companies cashed in on the movie's anticipated success through the design and distribution of *Jurassic Park*–inspired toys and sleeping bags and a theme-park ride at Universal Studios in Florida. "If dinosaurs had been marketed half as well as they're going to be in Steven Spielberg's $60 million-plus *Jurassic Park*," wrote Pat H. Broeke in *Entertainment Weekly*, "they would never have become extinct."[15] Indeed, $65 million was spent in marketing to match the cost of the film's production. In the United States, the film generated more than $3 million from the June 10 midnight screening alone. After that night, the opening weekend generated another $47 million in sales. The film was also an international success, breaking opening records in the United Kingdom and several other countries from Japan to Taiwan. Overall, Spielberg's *Jurassic Park* earned more than $914 million worldwide in its first run, making it the highest-grossing film of the year. It soon became the highest-grossing film of all time, surpassing *E.T.*—another Spielberg-directed film—which had held the title for a decade.[16] In addition to its record number of sales, *Jurassic Park* won three Academy Awards for Best Sound, Best Sound Effects, and Best Visual Effects, as well as over twenty other awards, including international ones.[17]

Jurassic Park's success on screen was in part a product of its realism. Indeed, its success stemmed from its computer-generated images (CGI), which brought long-dead creatures like dinosaurs to the screen and rendered them lifelike and entirely convincing.[18] With the rise of the Hollywood block-buster in the 1980s and 1990s, filmmakers wanted to turn far-fetched ideas into realistic images on screen, to make the fantastical appear believable.[19] As film studies scholar Michele Pierson wrote, "In the build-up to *Jurassic Park*'s release, speculation about the film's computer-generated dinosaurs generated by far and away the most publicity for the film." According to Pierson, *Jurassic Park* successfully achieved this when the first dinosaur, the giant, long-necked brachiosaur, appeared onscreen to the amazement of both the characters in the movie and the audience in theaters.[20] The innovation of CGI and its incorporation into *Jurassic Park* was both a technical and an aesthetic achievement.

Vision, along with the ability to achieve it, was another essential component to the film's overwhelming fame. Crichton, Spielberg, and the film production team, for example, made a scientific fantasy look like reality. "There was no one like Crichton, because he could both entertain and educate," his agent Lynn Nesbit later said in an interview with the *Los Angeles Times*. "His brilliance was indisputable, and he had a grasp of so many subjects—from art to science to technology." Spielberg said that Crichton's "talent out-scaled even his own dinosaurs," and that "he was the greatest at blending science with big theatrical concepts, which is what gave credibility to dinosaurs again walking the earth."[21] Spielberg himself had a similar reputation for delivering highly imaginative but borderline reality films as the director of the box-office hits *Jaws* and *Indiana Jones*.[22] Together, Crichton's and Spielberg's delivery of great entertainment made *Jurassic Park* a global success.

There were other reasons behind the worldwide fame of *Jurassic Park*—namely timing and scientific plausibility. In a four-page *Newsweek* article—"Here Come the DNAsaurs"—Sharon Begley noted that the popularity of the book and movie depended on timing: "All great science fiction must be science first and fiction second. Even more, it must tap into the reigning scientific paradigm of its era. For Mary Shelley's 'Frankenstein,' that paradigm was electricity. . . . For Godzilla, it was radioactivity and the Bomb. For 'Jurassic Park,' it is biotechnology." In this media report, Begley quoted Crichton: "'Biotechnology and genetic engineering are very powerful,' he says. 'The film suggests that [science's] control of nature is elusive. And just as war is too important to leave to the generals, science is too important to leave to scientists. Everyone needs to be attentive.'" In addition to timing, the film's plausibility rested on the science behind it. "This movie depends on credibility, not just the special effects," Spielberg told *Newsweek*. "The credibility of the premise—that dinosaurs could come back to life through cloning of the DNA found in prehistoric mosquitoes trapped in amber—is what allowed the movie to be made."[23] The fact that the book and film were based on the science of ancient DNA research made the idea of bringing dinosaurs back to life via DNA from insects in amber seem theoretically possible, if not imminently achievable.

This intimate connection between the science and science fiction of ancient DNA research offered scientists opportunities for publicity that many could, and in this case did, use to their advantage to promote their research and enhance their reputations.[24] According to the *Los Angeles Times*, Hendrik Poinar set up shop in the movie theater lobby on the opening weekend of *Jurassic Park*, selling bits and pieces of amber in hopes of raising funding for future research: "'Step right up,' barked the stocky, fresh-faced young man in a polka-dot tie. 'Step right up and see the real science.'" Meanwhile, Cano was bombarded by media reporters. "'And what they all really want me to say,' Cano said, 'is that this is possible, that we can clone dinosaurs.'" The *Los Angeles Times* wrote, "Unfortunately, he explained, this cannot be done now, will not be done ever and, even if it could be done, probably should not be done—for a whole host of moral, ethical and practical reasons. But why spoil a good story?"[25] In this case, these scientists saw *Jurassic Park*'s celebrity as a chance to promote their own image, as well as the overall image of ancient DNA research, by placing their work front and center with the movie's release and all the publicity that came with it.[26] However, they were not the only ones to do so.

THE RACE FOR DINOSAUR DNA

As the *Jurassic Park* hype continued, other scientists and scientific institutions capitalized on the film's fame to influence visibility and research funding. John R. Horner, a now famous paleontologist, started off studying geology and zoology at Montana State University in the late 1960s and early 1970s but ultimately never completed his degree. Throughout life, he struggled with undiagnosed dyslexia, hampering his educational experiences. Yet despite not having been awarded a formal university degree, Horner became a maverick fossil hunter, best known for his rare dinosaur fossil discoveries and controversial hypotheses regarding dinosaurian behavior. In the late 1980s, he was awarded an honorary doctorate from Pennsylvania State University and later achieved worldwide fame for his role as a scientific adviser to Spielberg on the *Jurassic Park* movie.[27]

In 1993, Horner proposed a project to the NSF to search for DNA in dinosaur bone, chiefly inspired by an unexpected finding that Mary Schweitzer, then a graduate student working with Horner at Montana State University and the Museum of the Rockies, had made when analyzing a bone fragment in the lab. Under the microscope, Schweitzer observed several unusual structures in a thin section of fossil bone from *Tyrannosaurus rex*. Small and round in appearance, these structures looked like red blood cells. After millions of years, the soft tissue structures should have decayed and been replaced by minerals during the fossilization process. But this bone fragment did not appear to be fully mineralized. "I got goose bumps," said Schweitzer in a media interview. "It was exactly like looking at a slice of modern bone. But, of course, I couldn't believe it." In fact, the bone seemed to contain organic matter, like red blood cells, which led Schweitzer to wonder if perhaps proteins or DNA might be preserved too.[28]

Horner and colleagues quickly applied to the NSF for a small grant—requesting approximately $35,000 for a two-year research project titled "An Attempt to Extract DNA from a Cretaceous Dinosaur *Tyrannosaurus rex*."[29] NSF funded the grant the same summer the *Jurassic Park* movie was released. According to one of the practitioners involved, the correlation between the funds and the film was no coincidence. "It's hard to get money. I think NSF gave us money at that time just because of the movie," they explained. "It was the perfect time for it" (Interviewee 16). Not only did NSF fund Horner and Schweitzer's research but they also scheduled a press release to coincide with *Jurassic Park*'s opening weekend.[30] The *New York*

Times covered the story, quoting a representative from the NSF who confirmed the deliberate timing of the press release with the movie premiere: "We thought it would be a good opportunity to get the word out on 4 of the 10 dinosaur research projects the N.S.F. is funding this year, including that of Mr. Horner."[31] In the end, Horner and Schweitzer obtained very tiny amounts of DNA from the bone, but they were unable to confirm the identity of the DNA as being of authentic dinosaurian origin. "It's easy to extract DNA from a dinosaur bone," Horner told a media reporter. "It's trying to prove that it's from the dinosaur—and not from some contaminant—that's hard."[32]

The search for the first and the oldest DNA, specifically dinosaur DNA, was drawing a crowd of interested practitioners. In a media news article for *Science*—"Dino DNA: The Hunt and Hype"—Virginia Morell noted that Horner and Schweitzer's work had "set off a furious race with other labs to be the first to publish on dinosaur genes." Cano, for example, had claimed to have extracted small quantities of DNA from dinosaur bone during an experiment earlier that year. However, he also noted they could not definitively determine whether this DNA was of dinosaurian origin or a contaminate from another organism.[33] Meanwhile, other scientists were on the hunt for more stable molecules, such as proteins, that might be preserved in dinosaur fossil material. Gerard Muyzer, along with colleagues from Leiden University such as Matthew Collins and Peter Westbroek (early practitioners in the study of fossil molecules), reported the recovery of proteins from several dinosaur fossils, including hadrosaurs and ceratopsians.[34] They claimed they could identify the exact protein, osteocalcin, but could not isolate it for further study. Around the same time, other scientists reported the identification of proteins from a sauropod vertebra believed to be approximately 150 million years old.[35]

Although the race for the first and most ancient DNA from the most iconic species, be it amber insects or dinosaur fossils, made for a good story, some scientists worried that the disproportionate attention the press and public gave to such studies would distort their view of the field. "Several groups are racing to get the first DNA out of dinosaur bones," reported Morell, "but other researchers say their efforts are taking attention away from the real scientific value of ancient DNA." Robert Wayne, an evolutionary biologist and current editor of the *Ancient DNA Newsletter,* feared that hype for multimillion-year-old DNA would overshadow the less sensational but more credible and scientifically significant ancient DNA analyses: "It's the new Disco Science question: Who's going to be the first to get dinosaur DNA?" According to Wayne, this new "Disco Science" was all about the first, the oldest, the most extraordinary, and the most unbelievable findings: "But the trouble is that these very topical questions tend to obscure other research on more recent materials—such as mammal pelts in museum collections—which are much more likely to contain real DNA from the original source."[36] The hunt for dinosaur DNA brought attention, as well as competition and tension, for a growing group of ancient DNA researchers.

The hype was not left unchecked. Above all, a number of ancient DNA researchers doubted that molecules, be it DNA or even proteins, could survive intact or unaltered for millions of years. Morell quoted Rebecca Cann, a former graduate student of Allan Wilson's at Berkeley and now one of the foremost geneticists of human evolution, who spoke to the studies claiming to have recovered multimillion-year-old DNA: "It's nasty, damaged stuff. We know from chemical experiments that it degrades and how fast it degrades. After 25 million years, there shouldn't be any DNA left at all."[37] Exogenous DNA (that is, environmental, bacterial, or human DNA) could be easily introduced to a specimen over time or through human handling in a museum collection or lab. This DNA, much more recent in origin, was often more intact and easier for PCR to isolate and amplify, therefore generating erroneous results. Consequently, practitioners demanded convincing evidence of ancient DNA authenticity. In a different news article, Morell quoted Noreen Tuross, a well-known molecular biologist at the Smithsonian Institution, who echoed this concern: "Given what we know about the decay of the DNA molecule, the onus remains on those who are searching for dinosaur DNA to prove that they've found it."[38]

In 1993, the Second International Ancient DNA Conference was held at the Smithsonian Institution in Washington, D.C.[39] A *Science* article— "Going for the Old: Ancient DNA Draws a Crowd"—noted the disconnect between ancient DNA's media coverage and what scientists were mostly concerned about in the lab: "While rejuvenated celluloid dinosaurs have grabbed headlines this year, these scientists were more concerned with topics such as tracing ancient human populations and understanding how DNA can survive the millennia."[40] Indeed, this meeting focused on a quite different side of ancient DNA research. Over the course of three days, the conference mainly featured technical talks on the biochemistry of DNA in terms of oxidation and radiation damage, chemical modifications, as well as sampling, extraction, and amplification techniques. Scientists were interested in theoretical explanations for why specific sources, like amber or dentin of teeth, appeared to be better storehouses for DNA. Meanwhile, others were focused on showing how ancient DNA research was relevant to evolutionary biology in terms of testing hypotheses about human evolution, migration, and colonization. In addition to all of this, there was a host of publications on DNA obtained from ancient plants, exotic and extinct species like the moa, early humans, and even fossil feces.[41]

Despite attempts to shift the focus to other areas and applications of this new line of research, the search for dinosaur DNA was far from a lost cause for some scientists. Up until this point, researchers had claimed the extraction of multimillion-year-old DNA from amber insects but actual evidence of DNA of dinosaurian origin had yet to be discovered. In 1994, *Science* published a study that seemed to change that. In this article, a U.S. research team reported the discovery of 80-million-year-old DNA from a bone fragment found in a coal mine in Utah. Although the scientists involved in the study were careful not to claim that the bone or DNA from it was dinosaurian in origin, their article did subtly suggest it: "On the basis of the circumstantial physical and geologic evidence, it is likely that the bone fragments belong to a Cretaceous period dinosaur or dinosaurs."[42] Further, the authors did not shy away from the suggestion when discussing their discovery with the media. "We called it a Cretaceous period bone. We never called it a dinosaur bone," said one interviewee who worked on the project. "I mean, I didn't necessarily *stop* anybody—[laughs]—from saying that or anything like that" (Interviewee 50). The *Los Angeles Times,* for example, reported, "Bone Yields Dinosaur DNA, Scientists Believe," while a *New York Times* headline read, "A Scientist Says He Has Isolated Dinosaur DNA."[43] *Science News* covered the story too with the headline, "Dinosaur DNA: Is the Race Finally Over?" Indeed, it seemed these scientists had won the race, but as far as Scott R. Woodward—first author on the article—was concerned, the hunt was far from over. Woodward's message was loud and clear: "Yes, you can get DNA from 80-million-year-old bones. This is just the beginning."[44]

INTERPLAY BETWEEN SCIENCE AND MEDIA

In the early 1990s, the search for DNA from fossils evolved under the influence of intense press and public interest, especially as it coincided with, then was thrown into the media spotlight by, *Jurassic Park.* This interplay between science and science fiction ultimately influenced professional interest, research agendas, publication timing, grant funding, and media coverage. As an international best-seller and multimillion-dollar blockbuster movie, *Jurassic Park* introduced ancient DNA research into both the professional and popular consciousness, raising awareness, attention, and even expectations for the evolving discipline.

Although the book's publication in 1990 garnered much notice, publicity peaked with the film release in 1993. For this already public-facing practice, the movie became the ultimate illustration in the minds of the media and the public of what science might one day accomplish. According to a younger but leading practitioner in the field, *Jurassic Park* became a "symbol" to help explain the science to the public and inspire scientists to become interested in it (Interviewee 12). The film, recalled another researcher, was "good press" for the field, ultimately inspiring a new and young generation of "geeky" but "glamorous" scientists (Interviewee 4). Another interviewee mentioned being inspired to get involved in ancient DNA research specifically because of the book and the movie. "I didn't want to make a dinosaur, but labs looked cool," said this scientist. "Ancient DNA sounds cool; sounds like it should be cool. Part of that really does stem back to *Jurassic Park*. It is still the legacy of that. That's when it entered the popular consciousness" (Interviewee 2). In many ways, *Jurassic Park* was synonymous with ancient DNA research: "The media think about *Jurassic Park* when they think about ancient DNA" (Interviewee 23). As science studies scholar Amy Fletcher notes, the movie served as a "cross-cultural metaphor for public discussion of ancient DNA research."[45]

The science studies scholar David A. Kirby echoed this point, explaining that although cinema often functions as an "alternative" and "informal" form of "science communication," it should never be regarded as "insignificant." In fact, film has had a significant effect on the public perception of science and technology. It can bring even the most fantastical sides of research to life. According to Kirby, "Film's reality effect renders scientific representations plausible because it naturalizes images and events within the fictionalized world." This realism can make the impossible seem possible, convincing viewers that what they see on screen is a real representation of the natural world: "Cinema is a powerful medium of communication because its reality effect provides it with a capacity to serve as a virtual witnessing technology. The more cinematic technologies advance, the better cinema becomes in serving as a virtual witnessing technology." Movies serve not only to help the public visualize and understand a complex scientific or technological idea but to legitimize the science too. It is equally important to realize that the naturalization of science and technology can have wide-ranging effects because, Kirby notes, "cinema rarely exists as a solitary entity," and "one need only look at *Jurassic Park* in its incarnations in novels, films, comic books, and computer games as well as its incorporation into television documentaries and news articles to see the high degree of intertextuality in science-based media." Movie science can take on "a life of its own outside the confines of the screen."[46]

Jurassic Park's fame and its link to the real science and technology of ancient DNA research elicited press and public enthusiasm, but what was notable was how researchers responded to the growing celebrity of their science. It was more than the media that cultivated this connection; researchers and research institutions were active participants. Some scientists cultivated celebrity because it translated into publicity and, in some instances, publication in high-ranking journals and research funding. Grimaldi, DeSalle, and the AMNH in New York capitalized on the book and movie's popularity to raise their public profile and bring awareness to the search for DNA from fossils; Raúl Cano and colleagues' publication in *Nature* on the recovery of multimillion-year-old DNA from amber insects aligned with the movie's release; Mary Schweitzer and Jack Horner's research proposal on the search for dinosaur DNA was funded by the NSF in the wake of the movie's success. Such deliberate and strategic efforts were quite practical and even beneficial.

This type of science-media interchange was not a one-off occurrence. Indeed, Kirby—in his work on the relationship between science and cinema—notes that scientists and scientific institutions often take advantage of working with or alongside the production of major blockbuster movies.[47] Specifically, he mentions that professional publications being "coincidental" with major movies is "common," pointing to the publication of Cano and colleagues' research along with other cases, such as the *Proceedings of the National Academy of Sciences'* publication of Mervin Herndon's controversial nuclear planet theory, which corresponded with the opening weekend of *The Core*.[48] More specifically, Kirby examines the ways scientists have served as "science consultants" on Hollywood movies. Kirby, for example, discusses Horner's role as a science consultant for *Jurassic Park,* as well as its sequels, detailing how Horner used his position to both influence public perceptions of dinosaur behavior and receive a "consulting salary" and "generous research grants" to finance his paleontology research.[49] There were obvious incentives for scientists to seek the spotlight and align their work with some of Hollywood's most memorable movies.

Although there were incentives to working in the media spotlight, there were also—at least according to scientists—disadvantages. "In one day, we must have spoken with 200 different journalists. . . . It was just amazing!" recalled a researcher involved in one of the early amber studies. "The media ate it up *big time* because *Jurassic Park* was out and the movie was just coming out." For this interviewee, there was a marked difference in the media's interest in this work on amber fossils. Earlier, this researcher had minimal interaction with the press: "The work that I was doing was not particularly interesting to anybody. . . . I never did anything that was earth-shaking. . . . The work that I did was *good* work, but it was a non-issue from the point of the media." As far as they were concerned, the difference between the former and the latter was a difference in news value: "That's the difference—work that you do that sells newspapers or gets airtime." At the same time, this researcher found the publicity "annoying" and even "troubling," as the media attention was quickly followed by a "pall of negativity" from the scientific community and "critical comments about the inability of DNA to last that long." The press and public eagerly accepted their research, but scientists were less receptive if not downright hostile. In the end, this researcher rather regretted the attention: "I had more than my fifteen minutes of fame" (Interviewee 31). Another scientist echoed a similar sentiment regarding heightened media attention, saying the publicity was overwhelming: "I was front page news *everywhere.* It was awful. Awful, awful, awful. Horrible!" For this researcher, the media attention and corresponding community competition, particularly in the early years, became reason enough for some scientists to opt out of the field: "I also learned that I don't want to work with DNA. Ever. Ancient DNA? Not ever" (Interviewee 39).

Crucially, the individual actions of these researchers and research institutions had a collective effect on the overall development of ancient DNA research as a discipline. In its early years, the ancient DNA community quickly recognized its role as a science in the spotlight. In the second *Ancient*

DNA Newsletter, circulated in 1992, Russell Higuchi addressed the increasing public interest around the idea of using DNA from insects in amber to bring dinosaurs back to life. For him, there was a time and a place for speculation, and as a scientist he believed that in certain contexts, too much unwarranted speculation did more harm than good: "When you get asked (and in the wake of *Jurassic Park,* the movie, it seems inevitable that some of you will) whether the resurrection of dinosaurs from ancient DNA is possible, I hope you will say it is not. Although it is fun to say, 'in theory, it may be possible (nudge, nudge—wink, wink),' let's get real." He admitted that this was easier said than done: "I myself have been guilty of allowing this romantic— if not gothic—notion, the resurrection of extinct species, to colour reports of our work (it *is* hard to keep the Media from focusing on that)." Nonetheless, Higuchi urged colleagues to find a balance between this close coupling of professional and popular expectations: "It now seems clear to me that the responsible thing to do is to try as much as possible not to overstate the power of new technology, in the field of ancient DNA or elsewhere."[50] Researchers were more than aware of the publicity their work garnered. They were also well aware of the need to balance the press's and public's expectations with the scientific and technological limitations.

THE ROLE OF CELEBRITY

Although the science and technology of ancient DNA research initially inspired *Jurassic Park,* the practice was in turn influenced by the celebrity that surrounded this blockbuster phenomenon. To some extent, *Jurassic Park* did actually drive and develop the hunt for DNA from fossils. In the early 1990s, the field emerged and evolved under the scrutiny of the press and the public gaze as a series of studies, published in respected journals such as *Science* and *Nature,* reported the recovery of multimillion-year-old DNA from amber insects and dinosaur bone. In the process, the press created opportunities for public exposure, but scientists fashioned their own occasions for attention too. The interplay between scientists and the media, specifically around the idea of discovering dinosaur DNA, influenced research agendas, publication decisions, grant funding, professional recruitment, visibility, and public perceptions of ancient DNA research. During this decade, some scientists were savvy in capitalizing on the celebrity of their rapidly evolving discipline in order to secure their success on both an individual and group level. They used celebrity to help shape ancient

DNA activity in terms of influencing the questions they asked, the funding they received, and the ways they framed their research when communicating to wider popular and political audiences about its significance.

For the most part, the pursuit of ancient DNA was followed by intense publicity in the form of consistent media interviews, newspaper reporting, magazine articles, and more. By the mid-1990s, however, the discipline had appeared to achieve a level of celebrity, a status of being well-known for being well-known, thanks to the global success of *Jurassic Park*. Ancient DNA research had become much more than a public-facing science. It had become a celebrity science.

This transition from a public-facing science into a celebrity science was very much a difference in the degree and duration of media attention projected onto the field. For example, although most scientific research and technological innovations enjoy publicity from time to time via headlines and special article features, or through characterizations in science-fiction and nonfiction books and films, not all science falls subject to celebrity. In other words, all celebrity involves publicity but not all publicity leads to celebrity. Celebrity is much more than intermittent promotion and attention, and the field of ancient DNA research was both witness to and evidence of this.

Celebrity facilitated ancient DNA's disciplinary formation in terms of community cohesion and identity. In this early and exploratory phase of research, there was not necessarily a unifying conceptual or theoretical framework for the study of DNA from ancient and extinct organisms. Yet speculation around the theoretical preservation and potential extraction of DNA from fossils, and even the hypothetical resurrection of extinct species, created interest in the practice. The media more generally, and *Jurassic Park* specifically, helped give definition and direction to this emerging but evolving discipline. In other words, media attention (much like the early conferences, scientific publications, and newsletters on ancient DNA research) was influential to the young field's formation. Scientists were savvy in harnessing the press and public attention in their search for DNA from fossils to generate, then sustain, activity in the practice at one of its most speculative and vulnerable stages. Under the name of "ancient DNA research" and in the wave of hype around it, researchers were unified in their investigation of DNA from fossils even at a time when the practice lacked clear and consistent financial or institutional support.

Although much of the hype around the search for DNA from fossils was focused on *Jurassic Park* and its influence on the public's perceptions of the science, researchers were not in fact trying to clone dinosaurs. Rather, the majority of the community was interested in using DNA for the primary purpose of studying the evolutionary history of extinct and extant organisms, and testing hypotheses about evolution, variation, selection, and migrations of past populations. Even practitioners in pursuit of DNA from the days of the dinosaurs claimed that cloning was not their goal. "Big game DNA hunters keep pushing forward, and not just for the glory of being first," reported Virginia Morell. "Horner and Cano both say they want to use the genes to reconstruct dinosaur evolutionary history."[51] Nonetheless, the ability to recover DNA from fossils, and the possibility of discovering multi-million-year-old DNA, was nearly inseparable, certainly in the minds of the media and its public, from the idea of resurrecting extinct species. To be sure, the celebrity around the idea of extracting DNA from fossils and using that DNA to bring back extinct creatures played a key role in ancient DNA's disciplinary development during this decade.

CHAPTER FIVE

Imposing Limits

THE PCR POLICE

By the mid- to late 1990s, a number of extraordinary publications on the recovery of ancient DNA had attracted widespread suspicion from researchers within and even outside of the ancient DNA community. As the discipline developed, practitioners began to respond to the hype that had been building up around it. This took form as an enthusiasm and confidence in the ability of scientists, with the aid of PCR, to consistently and reliably sequence DNA from a range of fossils, then to use that genetic information to learn about species origins, evolution, and migrations across time and space. Hype also manifested as speculation concerning scientists' potential to one day use DNA to resurrect extinct creatures, including dinosaurs. In responding to hype in both its forms, researchers felt the need to address the field's technological challenges and status as a public-facing practice. In fact, several researchers took it on themselves to address the discipline's ever present contamination concerns. They also sought to push back against what they viewed as disproportionate or undeserved publicity. As they saw it, too much media coverage, especially directed at studies making exceptional but questionable claims, was a second source of contamination that could challenge the legitimacy of the science.

In 1993, Tomas Lindahl, a well-known expert in DNA decay and repair at the Imperial Cancer Research Fund in the United Kingdom, spoke out publicly on these contamination concerns. As a specialist in the molecular behavior of DNA, Lindahl was highly suspicious of recent evidence that seemed to suggest the extremely long-term preservation of DNA. Such findings defied all past and current research on the principles of DNA degradation. It seemed to defy the chemical composition of DNA itself. In two separate articles, both published in *Nature*, Lindahl made these points and more, highlighting that processes such as hydrolysis, a chemical reaction that occurs to break down the compounds of a substance when exposed to water, posed serious problems for the preservation of DNA in fossil material.[1] He argued that DNA's biochemistry could not support such longevity, and in the event that it could, he was deeply worried about contamination. To control for contamination, he suggested using negative controls, conducting appropriate chemical analyses, and reproducing the results. This last recommendation was especially important but difficult to achieve because it required conducting the experiment again, using the same sample or a separate one, and obtaining the exact same results.

For these reasons, Lindahl found the work on multimillion-year-old DNA—or "antediluvian DNA" as he called it—to be particularly problematic.[2] As far as he was concerned, the credibility of such studies depended on proof of the authenticity of the DNA itself and the reproducibility of results. He cited two studies that claimed to have recovered multimillion-year-old DNA from amber insects—one by David Grimaldi and colleagues in New York and the other by Raúl Cano, George Poinar, and Hendrik Poinar in California—which he felt did not have adequate evidence to support their bold claims. In one of his papers published in *Nature*, Lindahl directly insinuated that the rather outstanding results these scientists had obtained were very likely the product of contamination. "It is hardly surprising," noted Lindahl, "that insect-like DNA can be detected by PCR in experiments carried out in a department of entomology."[3] In a clearly open attack on both of these recent amber studies, George Poinar set out to defend his team's work. In a reply to Lindahl, also published in *Nature*, Poinar argued that this "'off the cuff'" comment about contamination was not only uninformed but misplaced. "In our experiments," he explained, "none of the extraction, amplification or sequencing was conducted in a department of entomology, or of botany for that matter."[4] Poinar stood by the authenticity of their findings, but regardless of his defense, the question of contamination was out in the open.

Not only did Lindahl think that such studies were problematic but he also believed they distracted attention from other research: "Recent claims of recovery of 100-million-year-old DNA have overshadowed the valuable and important studies on moderately ancient DNA." For Lindahl, the next step in this field must be a conservative one: "Rather than proceed spectacularly further and further back in time with anecdotal reports on single samples, using the notoriously contamination-sensitive PCR, I suggest that the next goal be a convincing report on the amplification of small DNA fragments, say, 100,000 years old."[5] From his view, it was a matter of establishing DNA authenticity through the defining of criteria and a matter of redirecting research toward more promising avenues of study, even if that meant redirecting research away from the media spotlight.

In fact, there were several practitioners who welcomed Lindahl's criticism concerning ancient DNA authenticity. According to his memoir, Svante Pääbo especially applauded Lindahl's use of the term "antediluvian DNA"—initially used as a pejorative form of ridicule of studies reporting the recovery of very ancient, prehistoric DNA. Pääbo later recalled that he and his own lab at the time at the University of Munich in Germany loved it and used it because they felt it accurately characterized a growing body of unreliable research in the discipline.[6] From their perspective, this rhetoric was one way to separate their own work from what they viewed as more sensational and less credible research.

Pääbo welcomed Lindahl to the conversation about contamination because he felt that many in the developing discipline of ancient DNA research were not paying close enough attention to it. As early as 1989, Pääbo, along with Russell Higuchi and Allan Wilson, suggested a short list of criteria, from control extracts and independent extracts to phylogenetic comparison, to be implemented in laboratories in order to circumvent contamination.[7] However, Pääbo and others were increasingly frustrated with some studies that did not take these criteria as seriously as they suggested. Over the previous five or so years, the search for DNA from ancient and extinct organisms had brought together scientists from disparate disciplines. Those interested in the new field brought different scientific and epistemic cultures with them. At the intersection of these diverse backgrounds, researchers were faced with the need to merge various disciplinary values in order to reliably recover DNA from fossils and apply the results to a range of biological or historical questions. For several researchers, including Pääbo, it was crucial to standardize the practice, and this, as he emphasized, required specialized knowledge and solid training in molecular biology. "It was a great help to have a respected scientist from outside the field point this out," Pääbo wrote in his memoir, "especially given my concern that the ancient DNA field tends to attract people without a firm background in molecular biology or biochemistry who, lured by the media attention that accompanies many ancient DNA results, simply apply the PCR to whatever old specimen they happen to be interested in." This, as far as Pääbo was concerned, was "'molecular biology without a license.'"[8] According to both Lindahl and Pääbo, experts in molecular biology themselves, ancient DNA authenticity was a prerequisite for credibility on which both the success of the field and the reputation of the practitioners promoting it depended.

As excited scientists joined the hunt for ancient DNA, Pääbo became just as much a regulator of the new field as he was a researcher in it. According to a former doctoral student in his lab, Pääbo and the researchers and students working with him took on the role of "the PCR police" (Interviewee 12). They embraced a much more critical and conservative attitude, particularly regarding the work of their peers. They did so through advocating for criteria and openly demonstrating evidence that others' results were the product of contamination.

DEBUNKING DINOSAUR DNA

While Lindahl, Pääbo, and some other scientists found the results of these amber studies problematic, their distrust of multimillion-year-old DNA hit new heights when Scott R. Woodward, a microbiologist at Brigham Young University in Utah, and colleagues announced the recovery of 80-million-year-old DNA from bone.[9] In a paper published in *Science* in 1994, Woodward and colleagues were careful not to claim that the bone or the DNA from it were dinosaurian in origin, but they did suggest it, and the media reinforced the idea with newspaper headlines reading "Bone Yields Dinosaur DNA, Scientists Believe," and "Scientist Says He Has Isolated Dinosaur DNA."[10]

The sensationalism quickly turned to skepticism as several independent studies called into question the authenticity of these results.[11] *Science,* the same journal that first published Woodward's study, published a report by science writer Ann Gibbons titled "Possible Dino DNA Find Is Greeted with

Skepticism," which covered several researchers' initial concerns and their evidence.[12] For example, S. Blair Hedges, a biologist at Pennsylvania State University, and Mary Schweitzer, the paleontologist at Montana State University, critiqued the study on the grounds that it lacked appropriate phylogenetic analyses and additional attempts to replicate results prior to publication. According to Woodward's team, they had recovered evidence of several mitochondrial DNA sequences and determined them to be just as distantly related to birds and reptiles as to mammals. However, when Hedges and Schweitzer conducted an independent investigation, their phylogenetic analyses suggested that the DNA sequences were not dinosaurian but mammalian in origin, thus likely the product of contamination, specifically human contamination.[13] Other studies suggested this conclusion too.[14]

Like Hedges and Schweitzer, Pääbo's new lab at the University of Munich, where he had been recently appointed professor, strongly suspected that the DNA recovered by Woodward's team was actually a contaminate. In fact, it was his lab that provided clear evidence for it.[15] Hans Zischler, a postdoctoral researcher at the time, took the lead on this case. After conducting phylogenetic analyses, he too found that the supposed dinosaur sequences were more closely related to mammals than to reptiles or birds, but it was still not clear exactly to whom or what the sequences belonged. They appeared to be mammalian, perhaps human, but it remained to be definitively demonstrated. Pääbo and his lab thought on this, particularly on the nature of mitochondrial DNA. They knew that sometimes segments of the mitochondria can be transferred for various reasons from the mitochondrion to the nucleus of a cell, resulting in a special sequence referred to as a nuclear mitochondrial DNA segment. Pääbo's lab hypothesized that Woodward's lab had extracted a nuclear version of mitochondrial DNA with an unusual mutation, thus explaining the unidentifiable sequence (the supposed dinosaur sequence). To test this, Pääbo's lab devised a clever, and rather bizarre, experiment. Human DNA contains a mix of mitochondrial DNA and nuclear DNA sequences, the former inherited from the mother and the latter from the father. What they needed, explained Pääbo in his own record of events, was nuclear DNA, and nuclear DNA alone. The one way to get it would be from male sperm. As Pääbo recalled in his memoir, he asked his male graduate students to contribute to the cause by donating their own sperm from which Zischler would then isolate the head from tail to extract the nuclear DNA. He would sequence it for comparison with the suspected dinosaur sequence. In the end, they obtained a number of nuclear mito-chondrial DNA sequences from the sperm samples, two of which they found to be nearly the same as the suspected dinosaur sequence.[16]

They wrote up their findings and conclusions for publication in *Science* and, in a rather sarcastic fashion, tried to rationalize the striking similarity between their results and the supposed dinosaur sequences. First, they proposed that if Woodward's dinosaur DNA was in fact dinosaur DNA, then it must mean their sequences were similar because their own lab in Munich was actually contaminated with dinosaur DNA. They found this scenario highly unlikely. Second, they hypothesized that dinosaurs and mammals might have hybridized at some point before their extinction, therefore exchanging their DNA, thus explaining why the supposed dinosaur sequence looked more mammalian than dinosaurian. This too they found highly unlikely. Finally, they suggested that the extracts or equipment used in Woodward's lab were not clean but contaminated by human DNA. Quite obviously, Pääbo's lab found this conjecture most convincing: "In conclu-sion, these results strongly suggest that Woodward et al. accidentally ampli-fied nuclear copies of human mitochondrial DNA."[17]

Dinosaur DNA, in this case, was debunked. And the news of it, of course, played out publicly. A *Science News* headline read, "Dinosaur DNA Claim Dismissed as a Mistake."[18] A reporter for *New Scientist* wrote, "'Jurassic DNA' Looks Distinctly Human."[19] In an article for the *New York Times,* "Critics See Humbler Origin of 'Dinosaur' DNA," Malcolm Browne covered the story in detail. He discussed the issues of authenticity and reproducibility, as well as the caution that some scientists thought ought to be exhibited before making such big claims about truly ancient DNA. On this point, Browne quoted Hedges, who spoke to the caution he displayed in one of his own early experiments regarding the preservation and extraction of dinosaur DNA: "'We were working on an astonishingly well preserved fossil of a *Tyrannosaurus rex,* in which the bones themselves have survived without becoming mineralized into stone,' Dr. Hedges said. 'We found DNA sequences in the dinosaur bone that were in the right places and really looked like what we expected dinosaur DNA to be." According to Hedges, they submitted their results for publication in *Nature,* but when they found they could not replicate the results, they pulled the paper from publication. "Replication," as Hedges put it, "is one of the key essentials of scientific method."[20] Indeed, some scientists—especially those who played a role in demonstrating that the presumed dinosaur sequences were actually a case of contamination—believed that extraordinary claims should be accompanied by extraordinary evidence in order to prevent erroneous, and what many viewed as publicly embarrassing, results.

Clearly, the early studies on long-term preservation of DNA from amber insects fell into this category of extraordinary claims. And Pääbo's lab, not surprisingly, was one that questioned these studies and set out to test their validity. In 1994, Hendrik Poinar moved to Munich to join Pääbo's lab for his doctoral degree. While there, Poinar carried out research to try to understand the biochemical behavior of DNA and the particular environments in which it was more likely to preserve. Specifically, the goal was to create an independent experiment that could be used to confirm ancient DNA authenticity. In the lab, Poinar, Pääbo, and colleagues worked on a method called amino acid racemization, a test that used amino acids as biomarkers to determine DNA decay and its potential preservation in fossils. In an article entitled "Just How Old Is That DNA, Anyway?" science writer Robert F. Service explained the experiment: "An international team of researchers reports that a chemical change that converts amino acids in proteins from one mirror-image form to another—a process known as racemization—takes place at virtually the same rate as the degradation of DNA." Therefore, "if the amino acids show this conversion to even a modest degree, then the original DNA in the sample is likely long gone, suggesting that any remaining genetic material is a contaminant."[21] In other words, these scientists suggested that if the amino acids had been altered or could not be detected, then it could be safe to say that any accompanying DNA would also be altered or even nonexistent. Undamaged DNA would be a sign of recent DNA and therefore an indication of contamination.

In this specific study, the team tested the racemization of amino acids against the degradation of DNA using twenty-six different specimens ranging from several thousands to several millions of years old. Specimens included a mammoth as well as humans remains, dinosaur bones, and amber insects. For the most part, they found that multimillion-year-old specimens like dinosaur fossils showed extensive racemization. They even tested the fossil from Utah from which Woodward and his team claimed to have recovered DNA. This sample showed extensive racemization too. Indeed, other million-year-old organisms revealed similar results. They tested both fossil leaf remains and sediments from the Clarkia deposit in northern Idaho, the same place from where Edward Golenberg and his team claimed to have recovered the first evidence of multimillion-year-old DNA. Like the dinosaur fossils, these fossil leaf remains showed extensive racemization and therefore extensive DNA degradation. The results in the end suggested that retrieving DNA sequences from extremely ancient material was highly unlikely.[22]

However, there seemed to be an exception to the rule. In this same study, practitioners found that amber samples exhibited a lower level of race-mization. Indeed, they detected amino acids that seemed to be endogenous to the organism. Although they did not also detect any DNA, they hypothe-sized that amber resin, due to low water content, could provide conducive conditions for molecular preservation. Even Pääbo reasoned that the reten-tion of amino acids, and perhaps nucleic acids, could be attributed to the preservative properties of the resin itself.[23] Dinosaur DNA was out of the question, but based on these research results, the preservation of DNA from amber insects was open to debate.

In the wake of *Jurassic Park*, the United Kingdom's Natural Environment Research Council (NERC) funded the Ancient Biomolecules Initiative (ABI).[24] This initiative represented NERC's second funding strategy to expand on research that started with the Biomolecular Palaeontology Special Topic from 1988 to 1993. The ABI—chaired by chemist Geoffrey Eglinton at the University of Bristol and archeologist Martin Jones at Cambridge University—would fund three rounds of research projects to investigate the preservation and evolution of biomolecules across time from DNA to proteins, lipids, and carbohydrates.[25] According to several interviewees, the initiative was rumored to have been granted at that time in part because of *Jurassic Park* (Interviewees 9, 25, 46). One media article suggested as much: "The world-wide success of the film *Jurassic Park* has highlighted the need for projects funded by the Ancient Biomolecules Initiative." The piece continued, "Partly in response to the high profile that the film brought, the Natural Environment Research Council is providing the Ancient Biomolecules Initiative with about £2m for this area of research over a period of around five years."[26] Out of the three application rounds, the ABI granted money to twenty-one research projects, fifteen of which focused on ancient DNA from plant, animal, and human remains.[27] It also included funds to test the *Jurassic Park* hypothesis.[28]

The group that set out to test the *Jurassic Park* hypothesis—the reality of the long-term preservation and extraction of multimillion-year-old DNA from amber insects—was from the Natural History Museum (NHM) in London. In their application to the ABI, researchers outlined their objective to test the validity of such claims published in high-profile scientific journals against the hype. "Our work to date has shown that ancient DNA from four fossil insects, if it is present at all, is in exceedingly low copy number and/or highly degraded," they explained in their application. "We now wish to carry out the necessary detailed laboratory investigation to show, beyond reasonable doubt, whether any ancient DNA fragments exist whatsoever within amber-entombed fossils."[29] According to one researcher who worked on the project, there was a direct connection between this project and the film hype. "My job that I got at the Natural History Museum was all down to *Jurassic Park*," they explained. "The museum probably would never have got the funding to try and do this DNA from amber if it hadn't been for *Jurassic Park* in the first place. *Part* of my kind of entry into the ancient DNA world was all due to a movie; a fanciful fictional movie." Their objective, however, was not to recover dinosaur DNA and bring them back to life. "We weren't trying to get dinosaur DNA," clarified this interviewee. "We were trying to get insect DNA out of insects in amber" (Interviewee 25). Their project—"Ancient DNA and Amber-Entombed Insects: A Definitive Search"—was determined to produce a closed case either for or against the preservation and extraction of DNA from amber-preserved fossils.[30]

At the NHM, researchers including Richard Thomas, Andrew Smith, Richard Fortey, Andrew Ross, and Jeremy Austin undertook a comprehensive experiment to recover DNA from fifteen samples of amber insects, each from different resin types and time periods.[31] George and Hendrik Poinar supplied some of the samples, including the bee specimen that provided the initial evidence for the preservation and extraction of DNA from ancient amber.[32] After performing DNA extractions and PCR amplifications on the fifteen amber insects, the research group carried out negative controls and conducted phylogenetic analyses to check for ancient DNA authenticity. In reminiscing on the study, one researcher recalled how they "did all the protocols" and "jumped through all kinds of hoops" to ensure they had "believable results" (Interviewee 24). Despite such extensive experiments, however, they were unsuccessful in identifying any evidence of DNA from these multimillion-year-old insects in ancient amber.

In 1997, they published their findings in the *Proceedings of the Royal Society*, announcing that they had "singularly failed to recover authentic ancient DNA from amber fossils."[33] They explained that their results, coupled with previous research by two additional teams, provided convincing evidence that amber-preserved insects were not a consistent and reliable source of molecular preservation.[34] Their findings concluded, "The incompatibility between these and our negative results and previous reports of ancient DNA from amber-preserved insects is difficult to reconcile without suggesting some form of cryptic contamination in the latter." As far as these researchers were concerned, the existence of DNA from amber fossils was not much more than a "biological curiosity."[35]

Following years of speculation and even scientific evidence in support of it, the *Jurassic Park* hypothesis appeared debunked at last. "'No Go' for *Jurassic Park*–Style Dinos," reported *Science*.[36] "Lights Turning Red on Amber," stated *Nature*.[37] Although the scientists searching for multimillion-year-old DNA were only a small subset of the overall community, the whole of the field was affected by the high-profile nature of their claims and the equally high-profile responses to their refutation. This led to a dramatic drop in public and professional confidence in the field's credibility related to ancient DNA authenticity and the reproducibility of results.

The problem of contamination had come to the forefront just a few years earlier in 1995 at the Third International Ancient DNA Conference at Oxford University in England.[38] Together, Tomas Lindahl and Svante Pääbo tried to enforce professional and philosophical values of protocols and precision. According to one of the early leaders in the discipline, "Lindahl gave his talk about it being impossible for DNA to survive for too long, and Svante made a really eloquent talk about the need for rules and rigor within the field." Consequently, "everybody went away really impressed with the fact that we had to sort of self-regulate ourselves." As far as Lindhal and Pääbo were concerned, there was a lot at stake if scientists in this new field did anything otherwise. "And I think the message that Svante was trying to get across was that if we don't self-regulate ourselves then we will lose credibility and the field will completely die" (Interviewee 4). However, the need to self-regulate was not just a private plea to the community. It was a public one too. A *Science* report read, "But the hype—and the embarrassment when some claims did not hold up—is causing ancient DNA researchers to fear that their field won't be taken seriously."[39]

NOVELTY TO MATURITY

As early as 1992, the community of ancient DNA researchers was conscious of their tendency to search for DNA from specific specimens that would be sure to appeal to the public, as well as popular but prestigious research journals. Robert Wayne, then at the Zoological Society of London, and Alan Cooper, a doctoral student at Victoria University of Wellington in New Zealand, were the initial editors of the *Ancient DNA Newsletter,* and in one of the first issues they highlighted this phenomenon. Over the previous few years alone, more than 35 percent of ancient DNA papers had been published in high-profile and high-impact research journals such as *Nature, Science,* and *Proceedings of the National Academy of Sciences.* "However, before we bask in self-congratulatory splendor," Wayne and Cooper wrote, "we should realize many of the papers concern just a few samples of disco species." The disco species, according to their view, were the flashy and iconic multimillion-year-old specimens often featured in studies as being the first or the oldest DNA. "The novelty of ancient DNA will soon disappear, requiring that we address more fundamental evolutionary questions."[40] Even at the height of hype, researchers recognized that the excitement of the practice and its time as a science in the spotlight might dissipate sooner rather than later.

Amid reports for the first and the oldest DNA came a more conservative movement as researchers were encouraged to set their sights on less geologically ancient specimens. Lindahl was one of the earliest proponents for this move. In fact, he—along with Pääbo—had argued since the early 1990s for the defining of criteria and a refocusing on younger samples. As he saw it, the flashy, headline-grabbing studies on multimillion-year-old DNA ultimately detracted attention from other more relevant and valuable work in the field, often work conducted on less ancient but more scientifically interesting specimens. Here, Lindahl argued that instead of focusing efforts on reaching further back in time, practitioners should attempt to generate convincing evidence for authentic DNA 100,000 years old or younger.[41] In such specimens, DNA was both more likely to be preserved and scientists would be more likely to be able to extract enough of it for meaningful analysis. From his view, it was more important, if not arguably necessary, to establish reliable evidence for ancient DNA authenticity.

Although not all scientists heeded Lindahl's advice, there were a few that did. In 1994, two back-to-back publications in *Nature* reported the recovery of DNA from the extinct woolly mammoth. This is certainly a "disco species" of its own, and both studies drew on Lindahl's suggested criteria in hopes of demonstrating the authenticity of their results and, by extension, the broader possibility of reliably recovering DNA from specimens that were tens of thousands of years old. One study—by Pääbo and Matthias Höss at the University of Munich and in collaboration with Nikolai Vereshchagin at the Institute of Zoology in Saint Petersburg, Russia—claimed to have sequenced DNA from five mammoths ranging from 9,700–50,000 years of age.[42] The other study—led by Erika Hagelberg, Mark Thomas, and Charles Cook Jr. of the University of Cambridge, in collaboration with colleagues Andrel Sher, Gennady Baryshnikov, and Adrian Lister—recovered DNA from two mammoths, one of which was at least 47,000 years old and presumed to be the oldest DNA from a vertebrate to date.[43] Interestingly, these achievements came nearly two decades after Allan Wilson at Berkeley, arguably one of the founders of ancient DNA research as a field, had attempted to identify genetic information from a frozen baby mammoth carcass found in Russia. Although Wilson and his lab extracted evidence of ancient proteins, they were not successful in extracting and identifying DNA. Consequently, these two studies provided the first evidence for this type of preservation. Together, they suggested the reliable recovery of DNA from organisms that has been dead for thousands of years without the contamination concerns that plagued those studies claiming to have discovered multimillion-year-old DNA.

As researchers turned their sights to less geologically ancient specimens, the specimens they studied were not necessarily less newsworthy from either a public or professional perspective. Indeed, there were other studies that enjoyed a good deal of media exposure and that simultaneously validated the potential of obtaining and using ancient DNA data to answer some of evolutionary biology's most complex questions.[44] One of the best-known studies of this decade was none other than the recovery of DNA from our ancient and extinct cousin, the Neanderthal.[45]

Over a century ago, in the mid-1800s, the first Neanderthal specimen was found in a cave in the Neander Valley, Germany.[46] At the time, no one knew this would become one of the first and most famous Neanderthal specimens in the history of paleoanthropology, and it was not until ten years after its discovery that researchers formally recognized the skeleton as such and declared it a separate species. This discovery, among others throughout the nineteenth and twentieth centuries, garnered extensive scientific and public interest around the study of human origins.[47] Since then, the Neanderthal's place in human history—and its extinction nearly forty thousand years ago—has provoked heated debates among scientists and the public alike. Up until the late twentieth century, however, the Neanderthal's relationship to modern humans remained unresolved.[48] Now, a century or so after the original Neanderthal specimen was found, Ralf Schmitz—curator at the Rheinisches Landesmuseum in Bonn, Germany, where the remains were housed—sought to change that. He knew there was work going on in the field of ancient DNA research that could potentially shed light on the Neanderthal's origins and evolution, and he thought that Pääbo, one of the leaders of the new field, might be up for the challenge.

After enlisting Pääbo's assistance, Schmitz provided a small sample of bone, which Matthias Krings, a graduate student conducting the study, then ground into a fine powder to extract DNA from it. Following a successful DNA extraction and PCR amplification, they were able to obtain 379 nucleotide base pairs of mitochondrial DNA, which they then compared to more than 2,000 mitochondrial DNA sequences from modern humans to determine its authenticity. What they found was good news. The Neanderthal DNA sequence varied from the modern human DNA sequences at an average of 27 positions. By comparison, modern humans differed by an average of just 7 positions. This meant that the Neanderthal DNA sequence was four times as different from the modern human DNA sequences and unique to the Neanderthal itself. The result was extraordinary, and they knew they needed further evidence in support of it. They needed to reproduce these results, preferably in a different lab independent of their own.

For the job of reproducing the results, they turned to Mark Stoneking, a geneticist specializing in human evolutionary history who had studied with Wilson at Berkeley for both his graduate and postdoctoral work in the 1980s. Stoneking, now a professor at Pennsylvania State University, had been involved in a number of landmark studies on the molecular evolution and origins of humans. One such study, led by Rebecca L. Cann and published in *Nature* in 1987, provided genetic evidence for the "Mitochondrial Eve" and "Out-of-Africa" hypotheses.[49] By comparing the mitochondrial DNA sequences of nearly 150 modern humans from across five regions of the world, they were able to determine their relatedness to one another and that they shared a maternal ancestor, popularly referred to as "Mitochondrial Eve." They also demonstrated that their common origin could be traced back to a single location, Africa. This lent support to the idea that humans first evolved in Africa, then migrated out of Africa to eventually populate the rest of the globe. The prospect of extracting genetic evidence from a Neanderthal, thought to be an extinct and distant cousin to living humans, was naturally exciting to Stoneking. In fact, it had been something of a dream for Wilson and his lab to use DNA to understand the relationship between modern humans and other early and extinct hominids such as Neanderthals.[50]

Stoneking agreed to attempt to replicate the results; a doctoral student, Anne Stone, who had spent a year in Pääbo's lab and was familiar with their protocols, took on the task. Here, Stone's goal was to extract, amplify, and sequence Neanderthal mitochondrial DNA that would be identical to select regions of the mitochondrial DNA sequences recovered by Pääbo's lab. She would attempt to achieve this by doing so in a different lab and by using a different piece of bone from the same Neanderthal specimen. The success of all this, of course, depended on precautions to minimize or avoid contamination. Unfortunately for these researchers, the first round of results was disappointing. Stone extracted and sequenced DNA, but it looked distinctly human, which obviously suggested that the equipment, reagents, or sample had been contaminated at some point in the process. If Neanderthal DNA were present, even in the smallest amounts, the PCR technique would have likely detected and amplified the contaminating DNA instead. According to Pääbo's memoir, he hoped this was the case and that there was still a chance to recover Neanderthal DNA. To test this, the lab came up with a new plan. Stone would use a different primer—a short, single-stranded DNA sequence used in the PCR technique to begin DNA synthesis—which would likely pair with the Neanderthal DNA, if it was indeed present, and not the modern DNA. Stone performed the study again, and this time the second round of results proved more promising.[51]

However, Stone was still not sure if the sequence from her lab matched the sequence from the Munich lab. The two teams would have to compare them to confirm their authenticity. Over an anxious phone call, they compared the differences in sequences one by one. "I called and read off the differences to [my colleague] on the phone," recollected one of the scientists involved. "I would say one and [my colleague] would say, 'Yay!'" (Interviewee 30). The reading of the sequence, as well as the celebratory responses to it, went on, one by one. With match after match, they concluded that both labs, independently of each other, had recovered authentic Neanderthal DNA.

In 1997, Krings and his colleagues from the Pääbo lab published their paper in *Cell* on the first evidence of Neanderthal DNA.[52] The publication was important for a number of reasons. Most obviously, the research shed light on the evolutionary relationship between humans and their Neanderthal ancestors. The researchers explained how they sequenced DNA, which they then compared with that of modern humans. Through such comparisons, they were looking for evidence of genetic similarity between the two, but based on the sequence data and subsequent analysis they found no evidence of genetic contribution. Specifically, they argued that Neanderthal mitochondrial DNA, when compared with mitochondrial DNA of primates and modern humans from Africa, Europe, Asia, and across the world, demonstrated significant differences, and they interpreted this as evidence that Neanderthals had lived, then died, without contributing any of their DNA to modern humans. In other words, they found no evidence that Neanderthals and our ancient human ancestors had interbred thousands of years ago. However, they also noted that this conclusion could not be definitively determined by mitochondrial DNA alone. These results also suggested that modern humans had their origins in Africa, not Europe. At the same time, the researchers reasoned that these findings did not completely rule out the possibility of a genetic contribution from extinct Neanderthals to extant humans; further data would be necessary to fully resolve this question. Nonetheless, this study and its attempt to use molecular data to inform a history that had traditionally relied exclusively on morphology added heat to an already virulent debate in evolutionary anthropology about our own origins and evolution over time.[53]

Further, this paper was noteworthy for where they chose to publish it. In his memoir, Pääbo recalled the reason for submitting this paper to *Cell* instead of to *Nature* or *Science,* as had been the tradition in the field of ancient DNA research: "Publication there would send a signal to the community that the sequencing of ancient DNA was solid molecular biology and not just about the productions of sexy but questionable results."[54] As far as Pääbo was concerned, a paper on ancient DNA research, published in a highly respected journal like *Cell,* but one that was in his view less about the headlines and more about the rigor, would work wonders for the field, especially at a time when its integrity was so openly contested. Pääbo wanted to show that their work was rigorous and relevant to evolutionary biology. Sure enough, this paper was intensely technical and methodological. Pääbo hoped to demonstrate that ancient DNA research was a serious business.

Despite Pääbo's decision to publish in *Cell,* the study did not escape press and public attention. Indeed, the celebrity that accompanied the science, the fossil itself, as well as the study's conclusions and implications for human evolutionary history were all reasons to expect an onslaught of media coverage. *Science,* for example, called it "a technical tour de force."[55] A report in *Nature* declared, "Given the quality of the new molecular findings, the study of ancient human DNA can, at long last, be said to be on a secure footing."[56] *The Guardian* of London spotlighted the research conclusions with the following headline: "We're African, No Bones About It."[57] Roger Lewin also published a piece in *New Scientist*—"Back from the Dead"— where he explored the research implications. Lewin quoted Tomas Lindahl, a notorious critic of ancient DNA activity, who said the study was a "landmark discovery" and "the greatest achievement so far in the field of ancient DNA research."[58] Ancient DNA always ran the risk of contamination, but for Lindahl this paper was "compelling and convincing" in its validity.[59] In light of these new results and the newfound confidence that scientists seemed to throw behind the science of ancient DNA research, Lewin brought back the idea of resurrection, saying, "We'll never resurrect dinosaurs but what about Neanderthals?"[60]

The recovery of Neanderthal DNA was a highlight of the decade, but it was also just one of many studies that addressed major questions in human evolutionary history, especially regarding the migration and admixture of past populations. Hagelberg, for example, had provided some of the earliest evidence for the fact that DNA could be preserved in and extracted from human bone.[61] Later, she and colleagues demonstrated the ability to amplify DNA from ancient human remains of Polynesia and then used this DNA to inform hypotheses about the migration and occupation of past peoples on these and other islands.[62] In the United Kingdom, Terence Brown and Keri Brown at the University of Manchester were instrumental in furthering the exploration of ancient DNA methods as applied to archeological specimens.[63] From this nexus, ancient DNA activity flourished as practitioners tried to learn more about human evolution, populations, migrations, diet, and disease as well as to determine the sex, ages, and kinships of past people. For example, researchers across the United Kingdom, France, Germany, Israel, and the United States began to investigate the preservation and extraction of DNA in ancient humans in terms of its application to questions in human evolutionary history.[64] Practitioners also published on evidence of leprosy and tuberculosis in ancient humans.[65] Together, these works seemed to suggest the utility of ancient DNA research as applied to archeological, anthropological, and epidemiological specimens, but contamination again was a worry.[66]

These concerns over contamination as it related to the study of ancient humans played out publicly. Here, the issue was not necessarily about whether DNA could last for several thousand years but whether the DNA that was preserved and extracted could be demonstrated to be authentic human DNA and not a contaminate. In this context, with humans working on ancient human remains, it was especially difficult to determine whether modern DNA could be contaminating whatever ancient DNA might be left in these samples.

In 1995, just a few years before the Neanderthal DNA discovery, Stoneking wrote an article for the *American Journal of Human Genetics*, "Ancient DNA: How Do You Know When You Have It and What Can You Do with It?" He wrote the article in response to both the professional development and popular attention that the field had achieved. As Stoneking noted, there were numerous articles on the topic, research funding opportunities, international conferences, newsletters, textbooks, and even rumors of a journal. In addition to all of this, there was "public notoriety," much of which centered around the book and movie *Jurassic Park*. Indeed, taken together, it seemed the search for DNA from fossils had, as Stoneking claimed, "'arrived' as a legitimate field of inquiry."[67] At the same time, however, he conceded there were more than just a few kinks to be worked out first. Sure enough, the field faced two troubling issues, namely the question of ancient DNA authenticity and the perceived value of ancient DNA data to broader research in evolutionary biology.

To address these issues, Stoneking considered a case by Elaine Béraud-Colomb of the Institute of Developmental Biology in Marseille, France, and colleagues who reportedly recovered DNA from several human specimens up to twelve thousand years of age.[68] He commended the "exhaustive procedures" the team took in order to control contamination, noting that the study seemed to meet the "informal guidelines" suggested by researchers like Pääbo. According to Stoneking, "This may seem like over kill to the uninitiated, but the ancient DNA community tends to be a rather

suspicious lot and likes to see some evidence that people are paying attention to the concerns that have been raised about avoiding contamination and
authenticating results." Although Stoneking agreed that independent replication in a separate lab was definitely preferable, he believed it was unrealistic. Here, he argued that it would be far from practical to make this kind of
replication a requirement for every study produced by every lab. In practice,
it would "cause more problems than it would solve" because independent
replication on this scale would be expensive, destructive, and ultimately
restrictive. He instead explained that "precautions" as well as "multiple independent extractions from each sample" would "suffice."[69]

In addition to issues of authenticity, Stoneking spoke to the utility of
ancient DNA data and particularly the novelty and celebrity of the field: "After
all, isn't it a neat enough trick to show that DNA can indeed be obtained from
ancient specimens?" In answering his own question, Stoneking wrote, "Alas,
if ancient DNA is to become a legitimate field of scientific inquiry, then the
answer must be no." He pointed out that recent research in the field,
including Béraud-Colomb and colleagues' work, simply showcased the
anomaly of ancient DNA from one or several samples with little insight or
impact into the larger looming questions in evolutionary biology. Stoneking
insisted that the authenticity and utility of ancient DNA research must
extend beyond its novelty: "If ancient DNA is to be more than a technological
curiosity, then we don't need any more such papers."[70] With increasing
interest in ancient human DNA came increasing issues regarding authenticity as well as utility. As an alternative to one-off extractions from supposed
"disco species," he recommended producing more sequences from many
samples to tackle anthropological questions on a population rather than individual level.

During this decade, researchers were responding to credibility concerns,
induced by both contamination and celebrity, by constructing criteria that
would help to transform the hunt for DNA from fossils from an emergent
into an established practice. Alan Cooper, now a postdoctoral researcher in
the Department of Biological Anthropology at Oxford University, was
becoming a dominant researcher in the field and a major critic of other
colleagues' studies. In 1997, he replied to Stoneking in the *American Journal
of Human Genetics*, reinforcing independent replication in light of ancient
DNA's short but sensational history: "Several ancient DNA 'triumphs' . . .
that have turned out to be embarrassingly unrepeatable, or contaminated,

might have been prevented if independent verification had been sought prior to publication." Cooper argued that adherence to hard-and-fast criteria ensured credibility: "In summary, there are currently several methods available to test the authenticity of ancient human DNA sequences. I suggest it is the responsibility of the ancient-DNA community, and archaeologists working with them, to insist that they are fully utilized. Failure to do so threatens the credibility of ancient-DNA research."[71] Ancient DNA's credibility was at stake, no doubt. As far as he was concerned, all studies must perform all tests and checks in order to demonstrate their reliability. Although Cooper and Stoneking agreed contamination was a problem, they differed in the extent to which criteria should be required.

After nearly a decade of headline-grabbing and contentious publications, the problem of contamination, one exacerbated by the celebrity that surrounded the science of ancient DNA research, began to divide the community. The field had made great strides, but its technical difficulties were more than apparent. Indeed, the community's credibility was on the line and some took the task of self-regulation into their own hands. This became more obvious at some of the later ancient DNA conferences toward the end of the decade. A biomolecular archeologist and early researcher in the field, for example, recalled these tensions: "I remember there was a conference where Alan [Cooper] was fourth or fifth speaker in a session and he was going to present some of his work. And he changed his talk. I saw him redoing his slides just before his talk. And he stood up and instead of talking about the work he was doing, he talked about how *rubbish* the field was and how human ancient DNA was becoming completely discredited." According to this interviewee, "Everybody hated him for it because he was just so rude. But it needed to be done. We'd just listened to four talks by people who . . . said, 'We've done this' and 'We've done that.' Is it actually genuinely true?" (Interviewee 4).

In addition to addressing the community at this conference, Cooper, along with Robert Wayne and Jennifer Leonard at the University of California, Berkeley, published publicly on the discipline's increasing professional and popular appeal: "From the beginning, ancient DNA research was a populist science. Reports of DNA from ancient remains led to wild speculation in the press and film that life could be restored to ancient creatures. Each new discovery served to reconfirm the public impression that scientists were moving quickly toward this goal. New reports of ancient DNA, although often of limited evolutionary significance, were published in the most prestigious journals." These researchers, among others, clearly realized that popular enthusiasm had been a major player in their early history. At the same time, they also believed that this interest alone was not enough: "The honeymoon period has passed for ancient DNA research, and the difficulties associated with a maturing field need confronting."[72] By the end of the decade, it was more than obvious that contamination was one of those difficulties. It was a call that the novelty of the practice must give way to its eventual maturity.

THE ROLE OF HYPE

Throughout the 1990s, a handful of practitioners had tested the limits of DNA preservation in what became a decade-long debate. This debate was initiated, and also very much kept alive, by a number of practitioners who sought to reach further back in time to the days of the dinosaurs in attempts to extract DNA from various specimens. This race for the oldest DNA, and in some cases dinosaur DNA, was influenced by the hype around Michael Crichton's book and Steven Spielberg's movie *Jurassic Park,* both released early in the decade. In light of this interplay between science and the media, interviewees characterized this decade and the search for DNA from fossils as the "Wild West" and even "the *Jurassic Park* phase" (Interviewees 10, 4).

By the end of the 1990s, however, contamination concerns as they related to ancient DNA authenticity placed the practice's credibility on the line. Here, the issue of contamination was illustrated most clearly, and most publicly, through those same research papers claiming to have extracted and sequenced multimillion-year-old DNA. Not long after these papers were published, other practitioners challenged the authenticity of their findings. In fact, some practitioners demonstrated that such bold claims were either irreproducible or the outright product of contamination. The overturning of these research results had devastating consequences for the community's reputation, so much that researchers had to work to establish their legitimacy in light of failed expectations. Crucially, the ancient DNA community was responding not just to these failures, but to the very public nature of those expectations. This was because the very studies that were demonstrated to either be irreproducible or the product of contamination had been published in high-impact journals such as *Nature* and *Science* and broadcast across mass media.[73]

Researchers felt a need to respond, to defend their credibility. In response to this, some scientists sought to impose limits by defining criteria to minimize contamination in the lab and by redirecting research toward more promising avenues of study on less geologically ancient specimens that would be more likely to preserve DNA and have preserved enough of it to be used for meaningful analyses. Lindahl, Pääbo, and to a certain extent Cooper and Stoneking were chief proponents of this conservative movement from the onset.

After nearly two decades of research into the long-term preservation of DNA from ancient remains, it was clear that the field faced severe technological limitations, resulting in community division and widespread skepticism in its potential to deliver on initial promises. Indeed, ancient DNA's disciplinary development had followed a similar trajectory to that of other scientific ideas and technological innovations. This trajectory—"the hype cycle"—characterizes the lifespan of an idea or innovation as a series of highs and lows in correspondence to the successful or unsuccessful attainment of expectations.[74] More specifically, the hype cycle is often described as moving from an initial trigger and peak of expectations followed by a trough of disillusionment, and at last a slope of enlightenment and a plateau of productivity. As science historian Elsbeth Bösl explicitly argues, the search for DNA from fossils followed this pattern of development.[75] In fact, a biomolecular archeologist who joined the search for ancient DNA early in the discipline's formation offered this observation too: "I'd say this research discipline has developed the way that all science—new scientific disciplines—develop in that you have an initial wonderful discovery, you have lots of hype and high expectations, and then you come down to it with a bump, and then you do the hard work of working out what it all means and what you can really do: what is realistic and what isn't" (Interviewee 5). According to this scientist, hype had become a driving feature in the search for DNA from fossils. Crucially, however, researchers were responding to more than just failed expectations as they related to limitations of the PCR technique and the longevity of DNA preservation.

As the new field neared the turn of the century, its practitioners found they had to work around and against two different but not unrelated problems affecting the credibility of the search for DNA from fossils: the problem of contamination and the problem of what they viewed to be too much undeserved or disproportionate publicity. In the first instance, hype took form as a confidence and enthusiasm, on behalf of both scientists and media reporters, in the ability to consistently and reliably sequence DNA from a range of fossils, then to use that genetic information to learn about species origins, evolution, and migrations across time and space. In the second instance, hype was projected as speculation concerning scientists' potential to one day use DNA to resurrect extinct creatures, including dinosaurs.

Hype, in both forms, was instrumental in the ancient DNA's disciplinary development.[76] In the early years, hype was performative in that it generated interest and guided activity that led scientists to experiment with novel ideas in order to establish evidence for their feasibility.[77] In the early stage of any innovation, the utility and reliability of the new technology or technique is not a given but must be demonstrated. "Hype corresponds to a particular phase in the career of innovations," explains sociologist of science Nik Brown. "The whole language of novelty, newness and revolutionary potential is actually part and parcel of the hyperbolic discourse surrounding the early or opening moments of resource and agenda building."[78] Indeed, it is often pragmatic, even necessary, to engage in hype to attract further attention from professional and public audiences alike in order to marshal interest and resources. In the case of the emergence of ancient DNA research, the press created opportunities for publicity but scientists also fashioned their own opportunities for attention. The interchange between scientists and the media, specifically around the idea of discovering dinosaur DNA, influenced publication timing, grant funding, research agendas, and professional recruitment. During this decade, some scientists were savvy in capitalizing on the celebrity of their fast-growing field in order to secure their success.

At the same time, hype—or too much of it, or not the right kind of it—posed a problem to the field's growth and acceptance as a credible and legitimate approach to studying evolutionary history. Brown captures this tension between hope and hype, explaining that while hype plays a role in generating activity and sustaining interest in new technologies and their applications, hype can lead to overshoot, which can result in damaged reputations. It is more than scientists' reputations at stake but the reputation of an entire research practice that can suffer from failed expectations: "In so many cases, the present fails to measure up to the expectations once held of it. This can have disastrous consequences for the reputation not only of individuals but entire innovation fields."[79] Sure enough, ancient DNA researchers felt that celebrity was a further form of contamination which could reduce the legitimacy of the new field.

The failed expectations of PCR technology as applied to very ancient fossil material were compounded by the intense public interest in the search for ancient DNA. In responding to hype, researchers felt the need to address the field's technological challenges and status as a public-facing practice. Specifically, a handful of researchers took it on themselves to address these ever present contamination worries. They also pushed back against what they viewed as disproportionate or undeserved publicity. The skepticism surrounding the ability to recover DNA from ancient human remains exacerbated these issues. As the discipline developed, contamination concerns challenged the authenticity of results. More broadly, these concerns deeply challenged the credibility of the practice. Given the discipline's short but sensational history, it seemed difficult, if not impossible, for scientists to move toward community consensus regarding ancient DNA authenticity. From enthusiasm to cynicism, ancient DNA research had emerged, evolved, and now struggled to become an established practice within evolutionary biology. The field's credibility was on the line, and according to scientists, contamination and celebrity were to blame. As a consequence, there was mounting distrust and discord among ancient DNA researchers. After two decades, three newsletters, and four conferences, the ancient DNA community was growing but clearly growing in different directions.

CHAPTER SIX

Contamination

DO IT RIGHT OR NOT AT ALL

In the summer of 2000, scientists convened for the Fifth International Ancient DNA Conference at the University of Manchester in England.[1] According to a meeting report published in *Science*, the discipline seemed to have emerged from its race for the oldest DNA and was entering a new phase of methodological development and scientific maturity. The author, Eric Stokstad, highlighted several studies that seemed to confirm this assessment.[2]

One study that Stokstad featured was by Svante Pääbo and Hendrik Poinar, at that time a doctoral researcher studying with Pääbo in Munich, whose recent research had found that coprolites, the scientific term for fossilized poop, were great sources of DNA.[3] In this study, they tried to extract DNA from approximately twenty-thousand-year-old fossilized feces found in a cave outside of Las Vegas, Nevada. Standard methods for extracting and amplifying DNA from modern fecal material were unsuccessful, so they used another approach involving a chemical compound, N-phenacylthiazolium bromide, which breaks down sugar-derived protein cross-links in order to release the DNA. By adding this compound to the extraction process, they were able to recover DNA and then determine, based on sequence comparison, that these coprolites were left behind by an ancient and extinct ground sloth, *Nothrotheriops shastensisthus*. From this data, scientists were given a glimpse into the diet of a long-lost species. Based on this study, caves seemed to be a unique site for molecular preservation, while the application of this method to fossilized feces revealed a unique source of molecular information.

Stokstad also spotlighted recent research led by Alex D. Greenwood and Ross D. E. MacPhee at the American Museum of Natural History, who claimed to have recovered the first evidence of nuclear DNA, not just mitochondrial DNA, from the extinct woolly mammoth.[4] In this same study, Greenwood and colleagues also claimed to have salvaged partial sequences of an endogenous retrovirus found in the DNA.[5] Stokstad explained that although endogenous retroviruses are common across all creatures and unlikely to offer information or insight about the life of a particular organism, this finding led the team to consider the possibility of tracing the evolution and extinction of species through ancient pathogens. Prior to this specific study, MacPhee and Preston A. Marx of the Aaron Diamond AIDS Research Center in New York City hypothesized that pathogens from encroaching humans or animals arriving in North America via the Bering Strait could be in part responsible for the mammoth's eventual extinction.[6] If viruses could be detected in the fossil record, then this could be the start of a new field dedicated to the search for, and the study of, paleoviruses. According to Stokstad, the scientists behind this research admitted that finding evidence of an Ice Age pathogen would come down, basically, to luck.[7] In highlighting new methods and sources of DNA, however, Stokstad portrayed the practice at the beginning of the twenty-first century as one ready to make some serious, cutting-edge contributions to our knowledge and understanding of evolutionary history.

A month after the meeting, however, *Science* published another article with a very different view on the state of the field. This article—"Ancient DNA: Do It Right or Not at All"—painted a picture of skepticism, criticism, and frustration with the discipline's development. The article was written by Alan Cooper—recent founder and director of the Henry Wellcome Ancient Biomolecules Lab at the University of Oxford—and Hendrik Poinar, recently graduated from Pääbo's lab in Munich and by this time a postdoctoral researcher at Pääbo's new lab at the Max Planck Institute for Evolutionary Anthropology (MPIEVA) in Leipzig. The publication was a direct response to the conference, and one presentation in particular that Cooper and Poinar thought had all too "boldly opened with the claim that the field was now mature and could move ahead with confidence." Indeed, Cooper and Poinar strongly disagreed with this assessment: "This optimism is unfounded," they argued, "as demonstrated by the notable absence of 'criteria of authenticity' from many presentations at the conference."[8] Although standards for ancient DNA authenticity had been suggested over the decade, and while some scientists had adopted them, Cooper and Poinar felt that many others still ignored them. The authors even noted that editors and reviewers of prestigious journals continued to publish work without proper controls that would ensure the findings were both authentic and reproducible. They blamed the collective scientific enterprise—scientists, editors, and reviewers alike—for failing to employ or enforce criteria to standardize the practice and restore its reputation as a credible discipline.

To combat contamination concerns, Cooper and Poinar proposed a list of nine rules for "criteria of authenticity." First, every study had to be conducted in a "physically isolated lab"—a lab specially dedicated to ancient DNA activity in order to circumvent any contamination from modern material. Drawing on guidelines outlined in previous publications, Cooper and Poinar argued that labs handling ancient material must be physically isolated from other molecular or microbial labs containing modern material.[9]

In the "Ancient DNA Lab," Cooper and Poinar recommended that "control amplifications," or multiple extractions, be performed to test for contamination. Other required controls included "quantitation" to first determine if enough DNA was available for amplification, as well as "cloning" to estimate the amount of endogenous DNA (DNA originating from the organism) in the PCR product. Further, they argued that sequences should show "appropriate molecular behavior" or evidence of degradation. In other words, ancient sequences were expected to be short sequences of less than five hundred base pairs. Longer sequences were assumed to be the result of contamination, or at least required justification to demonstrate that contamination had not occurred. Additionally, Cooper and Poinar suggested employing indirect evidence of DNA preservation via "biochemical preservation" of other molecules like amino acids. And in the case of human ancient DNA, where contamination concerns particularly ran rampant, they insisted that sequences should be extracted and analyzed from "associated remains" or from animal remains to confirm molecular preservation from the same environmental setting. They also argued for "reproducibility," namely that practitioners should be able to produce the exact same results from the initial extraction, as well as additional extractions, of the same specimen. Finally, Cooper and Poinar maintained that scientists should reproduce their findings through "independent replication." This meant that a second sample of the specimen should be taken, extracted, sequenced, and confirmed in an independent lab by independent researchers. Although they admitted that completing all nine criteria would be expensive and time-intensive, Cooper and Poinar believed doing so was vital to the future of the field.[10]

Other practitioners expanded on these newly proposed expectations, particularly regarding requirements for the ancient DNA lab.[11] For example, some scientists suggested that ancient DNA research be conducted in a physically isolated lab, ideally located in a building without a PCR lab, and fitted with specific ventilation systems with positive air pressure to prevent contamination via airflow when entering or exiting. Further, all equipment brought into the clean lab should be decontaminated with bleach or ultraviolet (UV) irradiation as appropriate. With each person's entry into the clean lab, researchers would be required to dress in full-body suits complete with gloves, shoe covers, hairnets, and face-masks to avoid contamination during experimentation. Afterward, UV lights are used to sterilize the space in preparation for the next use. Researchers were also advised to never enter the clean lab after working in the PCR lab to avoid cross-contamination between these workspaces. Ultimately, the physical separation of the ancient DNA lab from other labs, as well as the precautions that researchers are required to take when working in the lab, became a hallmark of the proper practice of ancient DNA activity. The presence or absence of a clean lab became a way in which some scientists measured the credibility of research results within this community.

Overall, Cooper and Poinar's "Ancient DNA: Do It Right or Not at All" was an influential though not isolated effort to address contamination concerns. Sure enough, their criteria of authenticity drew on various publications introduced over the years, some of which dated back to the late 1980s or early 1990s.[12] One much more recent paper published by Pääbo's lab in 1997 on Neanderthal DNA, for example, outlined in detail the methodological and technological precision that they thought necessary for demonstrating ancient DNA authenticity.[13] But still there was a sense of frustration on behalf of some scientists, including Cooper and

Poinar, who felt that others in the field were not taking contamination concerns and standardization seriously. One interviewee explained the frustration and subsequent publication this way: "Various people—like Svante [Pääbo] and Tomas Lindahl and Alan Cooper and Hendrik Poinar—said there are issues and most people listened to them, who were actually published in the early papers or the erroneous early papers. . . . But a few people didn't and carried on ignoring them and so on. Then, in frustration, these criteria get published" (Interviewee 6). For Cooper and Poinar, it was more than the credibility of just one or two or even a handful of studies at stake. Indeed, they believed the credibility of the discipline as a whole was in jeopardy. The one way to fix this, as far as they were concerned, was for fellow colleagues in the field to strictly adhere to the criteria of authenticity: "Failure to do so can only lead to an increasing number of dubious claims, which will bring the entire field into further disrepute." The future of the field depended on it: "If ancient DNA research is to progress and fulfill its potential as a fully-fledged area of evolutionary research, then it is essential that journal editors, reviewers, granting agencies, and researchers alike subscribe to criteria such as these for all ancient DNA research."[14] For Cooper and Poinar, enforcing criteria was necessary in terms of standardizing the field and establishing its credibility within evolutionary biology more broadly.

Several years after Cooper and Poinar published their criteria of authenticity, Pääbo and Poinar, along with other colleagues, reiterated the importance of guidelines, especially in terms of reproducibility and replication. In a hefty thirty-page review paper for the *Annual Review of Genetics,* they reflected on the field's nearly twenty-year-history.[15] Looking back to the early days, Pääbo and co-authors recalled the early studies in the mid-1980s that first claimed to have demonstrated the discovery of DNA from ancient and extinct organisms, namely the discovery of DNA from the quagga and the ancient Egyptian mummy.[16] But they also brought attention to the fact that the findings in these studies, the first of their kind, had not been successfully reproduced to guarantee authenticity. Pääbo later noted that even his initial research on mummy DNA likely contained modern contamination.[17] As Pääbo and co-authors argued, "They were in a sense precocious, since the amounts of DNA present in the old tissues were so small that the isolation of bacterial clones carrying the same DNA sequence was essentially impossible. The results could therefore not be repeated in order to verify their authenticity. Thus, the litmus test of experimental science—reproducibility—was hard or impossible to achieve."[18] In this article, the authors explained that the advent and adoption of PCR in the late 1980s to ancient DNA studies changed this by offering scientists opportunities to exponentially amplify the given DNA sequence under investigation. But at the same time, this ability to amplify and analyze more than a single sequence exposed evidence of two challenges for the search for DNA from fossils, namely sequence errors via molecular damage or modern contamination.[19] According to Pääbo and colleagues, two decades later, "Contamination remains the single most serious concern in the study of ancient DNA . . ., a reality reflected in the continuous evolution of techniques to avoid contamination as well as the addition to and modification of criteria of authenticity."[20]

In their review paper, Pääbo and colleagues listed the criteria of authenticity, insisting that following them was of "paramount importance" in order to uphold the discipline's credibility.[21] At the same time, however, these authors also argued that completion of the criteria alone was not enough and "cannot be taken as proof that a DNA sequence is genuinely ancient."[22] In fact, it was entirely possible that a result could be a false positive. They noted that if an organism is contaminated with a specific DNA sequence, in theory the criteria of authenticity may be satisfied but the result itself may still be inauthentic. To explain this point, they provided an example of a case where DNA was extracted from a nearly thirty-thousand-year-old bear tooth found in China. From this tooth, researchers were able to amplify human DNA sequences (sequences that were clearly not authentic in origin to the bear tooth), which they were then able to reproduce. In this case, the criteria of authenticity were satisfied but the sequence that was first produced, then subsequently reproduced, was contaminated from the start.[23] In light of this, the authors added that "the most important prerequisite for successful ancient DNA research is a highly skeptical attitude to one's own work."[24] The criteria, they explained, were more of a framework for conducting and validating research results. Scientific reasoning, or a critical attitude, toward the research must be a part of the process too. In other words, criteria were important, but criteria alone were not enough.

The extreme efforts that some scientists took to emphasize the importance of authenticity and reproducibility of ancient DNA analyses, not to mention the efforts they took to try to achieve both, was not unusual or unreasonable. Indeed, sociologists of science Harry M. Collins and Trevor J. Pinch,

for example, investigated the variety of ways in which a number of practitioners have depended on replication as a hallmark of successful experimentation in their own scientific endeavors. In addition to the *how*, Collins and
Pinch also investigated the reasons *why* these practitioners depended on
replication. Drawing on cases ranging from parapsychology to the detection
of gravitational waves and solar neutrinos, Collins and Pinch argued that for
many practitioners, replication of results—the ability to generate the exact
same finding again and again—was the cornerstone of scientific validity.[25]
Collins specifically explained that for the majority of practitioners, "reproducibility" was understood as evidence that what they were doing in the lab corresponded to reality and the "universality" of science.[26]

However, Collins and Pinch also noted that replication in practice is far
from straightforward. In fact, it is often a serious source of controversy. In
the early stages of a new research field, it is nearly impossible to know
whether the methods employed or results produced are indeed reliable. In
the beginning, the correctness of the outcome is open to debate, as are the
means for producing, then reproducing, that particular outcome, whatever
it may be. In one of his works, Collins put the point this way: "Usually,
successful practice of an experimental skill is evident in a successful
outcome to an experiment, but where the detection of a novel phenomenon
is in question, it is not clear what should count as a 'successful outcome'—
detection or non-detection of the phenomenon."[27]

In the case of ancient DNA research—even after nearly two decades of
exploration into the subject—the very presence or absence of DNA in fossil
material was subject to debate. Researchers were divided over both the theoretical existence of and potential evidence for ancient DNA. They were
divided over the *possibility* that DNA could exist intact for hundreds, thousands, or even millions of years in the first place, and they were divided over
the *proof* required to demonstrate that authentic and ancient DNA from
fossil material could be reliably as well as repeatedly recovered. Consequently,
the act of replication—initially intended as a standard of success—was
instead a source of never-ending disagreement.

Collins described this dilemma in terms of a kind of regress. Specifically,
he called this the "experimenters' regress"—a "paradox which arises for
those who want to use replication as a test of the truth of scientific knowledge claims."[28] In other words, "To know whether an experiment has been
well conducted, one needs to know whether it gives rise to the correct

outcome. But to know what the correct outcome is, one needs to do a well-conducted experiment."[29] And on and on the regress goes until scientists decide, by whatever means, on what qualifies as a correct outcome of a properly conducted experiment.

But overcoming this regress is hardly easy. Not only is the validity of methods or results vulnerable to disagreement but the actual ability to reproduce those same results with the same methods is often riddled with debate: "The problem is that, since experimentation is a matter of skillful practice, it can never be clear whether a second experiment has been done sufficiently well to count as a check on the result of a first. Some further test is needed to test the quality of the experiment—and so forth."[30] Therefore, even if researchers are satisfied with the first experiment, they may not be persuaded that the second experiment, initially intended to replicate the results, was carried out in a competent manner.

In theory and in practice, there can always be disputes over how an experiment was conducted. Replication of scientific experiments is an arduous business. For Collins, "experiments are difficult" because "much experimental skill is tacit." As Collins explained, tacit knowledge is unarticulated knowledge involving matters of intuition, judgment, and experience, all of which are affected by more tangible resources such as time and money. When it comes to experiments, tacit knowledge is both essential and elusive. Indeed, the very nature of experiments, how they are conducted and reported, can be the crux of the problem. "Therefore, an experiment cannot be fully described in print even with the best will in the world," argued Collins. "That the conventions of scientific writing and publishing prevent the promulgation of intimate details of experimental trials, errors, and fudges, only makes things worse. Thus, an experimenter who wishes to test another's findings by replicating the experiment finds it hard to know whether any failure to find the same result is a consequence of the result not being robust or the replication being unlike the original in some crucial way."[31]

PCR, although an invaluable technique that facilitated the growth of the field, was also the source of its many problems. Considering PCR's drawbacks in regard to contamination, a younger scientist explained, "It's a mess. Half the time nothing happens. If something does happen, you can't repeat it. You get contamination you can't get rid of and you have no idea where it's coming from." The entire process is like "voodoo in the lab" with "all these rituals," but in the end "you can't figure out where anything is coming from," this scientist said, laughing. "*It's so frustrating!*" (Interviewee 27). Even the Leipzig lab—a frontrunner in the world of ancient DNA research— struggled with technological limitations too. According to another researcher, a colleague at the MPIEVA in Leipzig warned against working in the field of ancient DNA research: "Whatever you do, don't get into this. It's a completely dead-end thing" (Interviewee 42). As time went on, and despite the huge strides that have been made in the field, many practitioners felt that one step forward almost always meant two steps back. "And we were kind of stuck," one recalled. "You had a whole generation of researchers— this is what they wanted to do—but they were bound up by the limits of the technology" (Interviewee 27).

The Fifth International Ancient DNA Conference was an obvious nexus of contention regarding contamination, and Cooper and Poinar's publication "Ancient DNA: Do It Right or Not at All" forcefully put the issue, as well as their solution to it, out in the open for researchers internal and external to the practice. Their criteria of authenticity became a hallmark of credibility, and the acts of reproducing and replicating results became measures of experimental expertise. For many practitioners, contamination was a highly technical issue, and Cooper and Poinar's criteria functioned as a gatekeeper in the process of determining the reliability of results.

DUAL CONTAMINATION

Criteria of authenticity were more than a response to the developing discipline's technical challenges. The criteria were also a response to its status as a public-facing science, or even arguably a celebrity science. To be clear, scientists were concerned about "contamination" in both a literal and figurative sense as it related to ancient DNA authenticity *and* the intense press and public interest that surrounded the science.

In the literal understanding of the term, contamination referred to a specimen's exposure to DNA from other sources such as the environment or bacteria. Human handling in the field, lab, or museum was a further problem that scientists had to be aware of and avoid. This was all complicated by the highly damaged nature of DNA from organisms hundreds to thousands of years old. Yet as this reality of contamination became more pervasive and more publicized, researchers became cognizant of a different but not entirely unrelated form of contamination. In the figurative understanding of the term, some scientists saw the media as having a negative influence. More specifically, they viewed the hype and growing celebrity of ancient DNA research, especially related to disproportionate or undeserved media attention, as something that was contaminating their credibility as a legitimate approach to studying evolutionary history. Such hype was instrumental in the search for ancient DNA research's emergence and evolution into a new scientific field. At the same time, however, the failed expectations, and the very public nature of them, became interpreted by those internal and external to the field as overshoot, which ultimately damaged the reputations of not just a handful of practitioners but the overall discipline.

Indeed, the turn of the century was a critical period for the field, and the Fifth International Ancient DNA Conference, as well as Cooper and Poinar's "Do It Right or Not at All" article, were the ultimate illustrations of the need for change and the impetus for it. It was a turning point as ancient DNA researchers were forced to face years of concerns regarding contamination, celebrity, and what some viewed as a general lack of regard for either. At the Ancient DNA Conference in Manchester in 2000, a scientist recalled that conference organizers assumed the audience would ask questions about the validity of results. "But nobody did," said this scientist. "Everybody was very British and polite and just sat there" (Interviewee 5). Another senior scientist offered a similar perspective: "We were at the lunch of an ancient DNA meeting bitching away about how most reports were repeating the same errors that we'd seen in the late '80s and in the mid '90s. It was just like, 'Ah, for fuck's sake, we're doing it again!'" (Interviewee 32). For this interviewee, contamination concerns coupled with the growing celebrity nature of the field were the reasons for their discontent, explaining that some scientists were "attracted by the sexiness of the work" and the opportunity of "publishing in *Science* or *Nature*" (Interviewee 32). As one practitioner explained, contamination and celebrity were also the motivating factors behind researchers' decisions to advocate criteria of authenticity: "So, this is why Cooper and Poinar published that paper. That's where it all came from and it's this lack of self-criticism . . ., standing back from your work and saying, 'Is this believable? Is this right?' I think most scientists do, but some don't because they just think, 'Wow, this is great! I've got a great result and I can get a paper in *Nature* or *Science*'" (Interviewee 5). "Do It Right or Not at All" was not just a call for criteria but a demand to discipline the search for DNA from fossils as it developed into a celebrity science.

Just as media attention empowered the young field, it seemed to undermine it, too. Sure enough, some researchers—working both in and outside the discipline—viewed the ever present attention around the search for DNA from fossils as a further and more figurative source of contamination. As far as one senior researcher was concerned, these reactions were tied to deeply seated worries about the discipline's corresponding credibility: "I think [it] was not just a defense against *Jurassic Park*," explained this interviewee. "It was a defense against the rest of the scientific community who were starting to look at ancient DNA as, like I said, a sort of charlatan type of research" (Interviewee 4). According to some scientists, too much media interest or influence in the face of failed expectations could be seriously damaging.

In a way, however, celebrity was embedded in the science and publication practices around it. Reflecting on the history of the ancient DNA community, for example, one geneticist presented this perspective: "I do think it was a community that was distorted by the probability of getting spectaculars. It was interested in getting spectaculars; famous fossil, bit of DNA, *Nature* or *Science*." Both scientists and scientific institutions, along with the media, played a part in encouraging and producing headlining research: "So, I think it was distorted by the attention that *Nature* and *Science* give, and maybe those journals deserve a little bit of criticism because they like the headlines and let some things in that weren't that scientifically interesting because they had the 'wow' factor." For this interviewee, the publicity associated with an exciting young discipline of study had an effect on the type of research personalities attracted to it: "If you work in a more mature field, the route to a big paper in *Nature* and *Science* is years of *painstaking* work. We had phases where you get your bone, do your PCR, sequence it, send your paper off—so it attracted and rewarded people who liked a shortcut to success, as papers in *Nature* and *Science* are perceived as a success." However, this interviewee also conceded having previously capitalized on the celebrity of the science by publishing a paper in *Nature* on a very small amount of mitochondrial DNA mainly because it was DNA from an iconic, newsworthy specimen: "I can't preach and I don't mean to be preaching, but I think it was a distortion" to publish minimal results in such a high-profile journal. When asked if it felt like this in retrospect, the interviewee replied, "Oh no. It felt like it then. We knew. So, in some ways it attracted some crazy dudes—some successful crazy dudes and some unsuccessful crazy dudes" (Interviewee 21).

In reaction to credibility concerns, ancient DNA researchers engaged in "boundary-work" as an attempt to separate their work from other research that they viewed as less credible. Thomas F. Gieryn, a science studies scholar, initially introduced this concept of boundary-work by drawing on a historical analysis of early natural philosophers' and scientists' struggles to achieve authority in the face of opposition.[32] In his complete treatment of the idea, Gieryn described boundary-work as a "sociological explanation for the cultural authority of science itself" via the "discursive attribution of selected qualities to scientists, scientific methods, and scientific claims for the purpose of drawing a rhetorical boundary between science and some less authoritative residual non-science."[33] According to Gieryn, there is no one way to do science but different ways to draw and redraw the boundaries of what we view as science.

As Gieryn aptly noted, scientists often engage in boundary-work when they feel their credibility, and by extension their authority, is under attack. When threatened by internal or external interests, scientists defend their research by drawing a line, or multiple lines, between their work and other activities—scientific or not—that they consider a challenge to their reputation. When this happens, boundary-work is a vital factor in the contest for credibility: "Boundary-work becomes a means of social control: as the borders get placed and policed, 'scientists' learn where they may not roam without transgressing the boundaries of legitimacy, and 'science' displays its ability to maintain monopoly over preferred norms of conduct."[34] Indeed, boundary-work is a process by which scientists continuously construct, deconstruct, and negotiate definitions of what counts as science.

In the history of ancient DNA research, practitioners felt the need to respond to credibility concerns on two fronts: in response to the literal worry regarding the contamination of fossil or genetic material in the lab and in response to the figurative worry regarding the contamination of their reputation through excessive press and public attention. In other words, ancient DNA researchers engaged in "double boundary-work" in which researchers build boundaries on two fronts in response to two different but not unrelated issues that appear to affect their autonomy, authority, and legitimacy.[35] In one sense, researchers responded to contamination concerns by building technical boundaries around the practice through the implementation of criteria via the use of certain technologies and techniques. Cooper and Poinar's "Do It Right or Not at All" article was an example of this on paper, while the presence or absence of an "Ancient DNA Lab" was a physical boundary used by some scientists to help demarcate reliable from less reliable work.

In another sense, they felt the need to respond to the celebrity status of the field by building boundaries through rhetorical strategies in an attempt to isolate their work from media interest or influence—something that scientists were starting to feel was contaminating the quality of the science. In light of contamination and celebrity concerns, researchers realized that the integrity of their work depended on a policing of it. For some this necessitated a public response. Rob DeSalle and David Lindley, for example, wrote *The Real Science of* Jurassic Park *and* The Lost World in an attempt to address the scientific versus fictitious aspects of *Jurassic Park* and its successful sequel, *The Lost World* (1997).[36] David Norman, a paleobiologist at the University of Cambridge, reviewed the book and claimed, "It debunks the whole scenario very effectively and is a perfect antidote to all the ridiculous hype surrounding these films. Steven Spielberg is a fantastically successful film-maker; he has created some of the best fantasy movies ever; and that is all that *Jurassic Park* and *The Lost World* are—pure fantasy, no more, no less."[37] Indeed, DeSalle and Lindley set out to draw a line between the movies and ancient DNA activity as a real scientific and technological practice. To be sure, their intent was directly motivated by the consistent publicity that surrounded the science of ancient DNA research. According to a practitioner involved in the early ancient DNA research studies, "That book comes directly from the media interest in ancient DNA work" (Interviewee 18). Other scientists responded similarly but through more conventional means. Adrian M. Lister, for example, a paleontologist and close collaborator with ancient DNA researchers, wrote an article called "Ancient DNA: Not Quite *Jurassic Park*," while Mary Schweitzer and science writer Tracy Staedter published a piece entitled "The Real *Jurassic Park*."[38] In both these works, the researchers alluded to the blockbuster movie but highlighted the fickle nature of molecular preservation and the predominant problem of contamination. Given these issues, among many others, they argued that the idea of bringing dinosaurs back to life was out of reach, at least for now, all in an effort to set the record straight.

DO IT WITH ME OR NOT AT ALL

While initially intended to reduce controversy regarding authenticity and reproducibility of findings, criteria of authenticity as articulated in the "Do It Right or Not at All" article effectively provoked further debate and division in the community. In the attempt to control contamination and celebrity concerns, some also tried to control competition by making it more difficult to participate in the practice, limiting access to the technique and to success in the field. Indeed, the criteria of authenticity, and the degree to which scientists observed or did not observe these standards, divided the community into different collaborations, conferences, and even places of publication. According to one of the earliest researchers in the field, "Ancient DNA: Do it Right or Not at All" was the ultimate embodiment of a new, conservative, and exclusive philosophy of how to conduct scientific research on ancient and extinct specimens. As this interviewee noted, a more apt title for this paper would have been "Do It with Me or Not at All" (Interviewee 11).

By the turn of the century, Pääbo and Cooper had become scientifically and politically powerful players in the field. Both had become directors of generously funded labs at top-tier research institutions. Their credentials and experience had earned them reputations as cutting-edge, even cut-throat pioneers in the search for DNA from fossils. Both researchers had started off as students with Allan Wilson in the late 1980s and early 1990s at Berkeley, the lab recognized as the birthplace of ancient DNA research. And early in their careers they were both published authors in the new field and vocal spokespersons on the direction in which the search for DNA from fossils should go. Inside the community, Pääbo became known as the "Dark Lord of Ancient DNA" (Interviewee 12) and Cooper as the "Chief Challenger" (Interviewee 28). In their own ways, they established their own schools of thought and schools of followers.

As prestigious and prolific centers of ancient DNA activity, Pääbo's and Cooper's labs exercised their authority over the future of the field, and their ideologies specifically influenced a younger generation of rising practitioners. Here, they were influential in the developing discipline of ancient DNA research, particularly in regard to their conservative stance on contamination issues, but they differed in their interactions with, or rather reactions to, the rest of the community. "I think they, together, had a pretty strong influence on this conservatism," recalled one of Pääbo's former students, "with the difference that Alan [Cooper] propagated more aggressively than Svante [Pääbo] did" (Interviewee 15). At Oxford, for example, one scientist recalled exposure to and interaction with the ancient DNA community: "I did my PhD with Alan, and being fairly naive and unprepared, I listened to what Alan was saying about everybody doing it wrong. And he would occasionally send me to conferences . . . and give talks basically about how people should be doing it properly, and that probably didn't start me off as being popular with people" (Interviewee 6). According to interviewees, Cooper's involvement with the community was much more active and assertive, while Pääbo took a different, but not less effective, stance toward research in the field.

At Leipzig, Pääbo—despite his role as a founder of the field—intentionally isolated himself, his work, and his lab from the rest of the community. His disassociation came through his absence at conferences and disregard for most, although not all, work outside of his lab. An earlier student of Pääbo's recalled the community schism through the deliberate negation and resulting ignorance of other research in the field: "It was present, definitely, but it took me some years to realize it because I started in Svante Pääbo's lab. . . . For some time, I didn't even realize that the other part of the community existed. . . . In Svante's world, it basically didn't exist. It was nothing one needed to cite, nothing one needed to read. So, I kind of knew there was something, but it was something completely unimportant" (Interviewee 15). Pääbo and Cooper, along with their schools of thought and followers, both promoted conservatism via hardline devotion to criteria of authenticity, but they used different strategies to separate their work from what they viewed as less credible efforts in the ever growing field of ancient DNA research. Together, this conservative philosophy shaped how their students, as well as colleagues and collaborators, viewed their own research in relation to the rest of community.

According to one of the early researchers in the field, ancient DNA appealed to practitioners across disparate disciplines, but as it did, it drew in a group of practitioners that, as far as this interviewee was concerned, were "amateurs," ill-equipped in the specifics and nuances of molecular biology. "The trouble with ancient DNA is that you get people thinking they can do it; people who were forensic scientists, people who were doctors, the sort of doctors who like to retro-diagnose what Mozart died from. [Laughs]" For this researcher, as well as others, the amateur interest in the field was a challenge to the ancient DNA community's credibility and the boundaries they tried to place around the practice. "So, you get these people who think they can do DNA and they don't have the right facilities or the right knowledge or the right understanding of ancient DNA" (Interviewee 5).

Concerns for contamination, or a lack thereof, ultimately divided members of the community. Reflecting on the conflict at this time, this researcher described the divide as embodied by "two different types of scientists: the ones who do proper work in laboratories in clean rooms and the other ones who work in forensic labs or even medical labs where there is no proper thinking about controls and contamination." Their differences in practices resulted in a "division" between what many in the community referred to as "believers" and "non-believers" (Interviewee 5). Specifically, this division centered around debates about contamination and scientific standards for avoiding it.[39]

While both sides of the schism were aware of contamination, these groups differed in the degree to which they employed certain methods to test for ancient DNA authenticity. Roughly, the non-believers were suspicious, even outright dismissive, of research produced by the believers. These non-believers—which included individuals such as Pääbo and Cooper, as well as their students and closest collaborators—viewed research by the believers as less rigorous and therefore less credible, less believable. Interestingly, these terms—believers and non-believers—were categories that scientists on both sides of the schism used in reference to themselves and others (Interviewees 6, 23, 28, 36). Some also referred to the schism as a difference between the "haves" and "have nots" (Interviewee 11). Although not all interviewees used both or even one set of terms to describe the split, they all recognized the split, though to differing degrees, and its influence on the sociology of their science.[40]

Most noticeably, this community schism took form by way of separate conferences, collaborations, and places of publication. At the Fifth International Ancient DNA Conference, the tension was tangible. One epidemiologist and self-subscribed believer recalled the event: "I went to this awful conference . . . and none of us were invited to talk, so we all had to have posters and they ignored the posters." As far as this interviewee was concerned, the overall conference "looked like a closed shop." Indeed, this scientist "came away with a very strong message—'Don't bother coming back!'" (Interviewee 23). This skepticism was more than obvious, to the point where some felt more than unwelcome. In response to this and other issues, believers and non-believers alike went their separate ways.

As a consequence of differences, the ancient DNA community schism grew into two different conferences: the International Conference on Ancient DNA and Associated Biomolecules (an extension of the original conference, primarily attended by the believers) and the International

Symposium for Biomolecular Archaeology (a new and distinct event for the non-believers). This split was first and most apparent in 2002, when the Sixth International Conference on Ancient DNA and Associated Biomolecules, a continuation of the early meetings, was hosted at the Hebrew University of Jerusalem in Israel.[41] According to a scientist and self-described non-believer, "There was a divide—there was a split—because you had the next ancient DNA conference going to Israel with people who published some very poor work on paleodisease DNA and so, you know, [my colleague] and I said, 'Well, we're not going to that one'" (Interviewee 5). Indeed, various researchers in epidemiology were interested in using genetic techniques to study the evolution of disease throughout history. Helen Donoghue and Mark Spigelman in the United Kingdom, as well as Susanne Hummel and Bernd Herrmann in Germany, for example, had been trying to extract evidence of ancient pathogens, such as tuberculosis and leprosy, from human skeletons and other remains.[42] Their work seemed to suggest the success of the ancient DNA techniques as applied to anthropological and epidemiological questions, but contamination, of course, was a concern. According to Cooper and colleagues, this body of research hinted at the "potential field of genetic paleopathology," but for them, this sort of work came with real reservations: "Unfortunately, some pathogens are common in the extant human population (e.g. *M[ycobacterium] tuberculosis*) or have congeneric relatives common in soil or other animal hosts (e.g. *Mycobacterium bovae*), raising the possibility of contamination through handling, burial, or the use of animal products."[43] In other words, there was more than one reason to be suspect of this type of work unless practitioners could demonstrate otherwise.

According to interviewees, those who identified as non-believers formed a different conference that they thought would better represent their desired epistemic standards for the practice of ancient DNA research. In 2004, this separate conference—the First International Symposium on Biomolecular Archaeology (ISBA)—was hosted at the University of Amsterdam as an alternative avenue for the non-believers who doubted the authenticity of work by the believers.[44] "The ISBA only started off because of the ancient DNA meetings" and "the split in the community," recalled one researcher and non-believer. According to this practitioner, the split came down to differences in perceptions of what really qualified as credible research: "Apparently, it's [the ancient DNA meetings] still going. And apparently, people are still publishing stuff about insane stuff which is completely wrong!" (Interviewee 22). Another scientist on the side of the believers also reflected on this schism. For this interviewee, the search for DNA from fossils was about the battle for and burden of proof. The 1990s was a "period of confirmation," and to a certain extent the believers feel their work is still being challenged today. "The bastards out there," laughed this scientist, "who say, 'We don't really believe'" (Interviewee 28). This schism was certainly acknowledged at the time but also expanded on and reinforced by interviewees' memories and retelling of it.

Although conferences were loci of community conflict, contamination concerns further divided the discipline into different places of publication. Another non-believer offered a similar story about how this all played out professionally. "There are two major divisions," explained this scientist. "You can see where they publish. The people on the more critical side [non-believers] tend to publish in higher-impact journals. The people who are less critical [believers] tend to publish in journals no one has ever heard of" (Interviewee 6). In particular, the believers claimed it was hard to publish through the more well-known outlets such as *Nature* and *Science*. Some said it was rumored that reviewers would reject or negate their findings when their research went under review. One scientist on the side of the believers said, "They [non-believers] threw the baby out with the bath water and they just disregarded any work that didn't have dedicated air-conditioned facilities for the sole use of ancient DNA work. They just negated anything that people like me found." This same scientist explained, "They would say, 'Well, of course you can't believe that because they don't do this and they don't do that.' It got very annoying." Given this perceived rejection or negation, the believers turned to alternative avenues for publication and recognition: "We just ignored them and published in medical microbiology journals or multidisciplinary journals" (Interviewee 23). The split, based on different groups of scientists from differing scientific traditions and of differing epistemic standards, ultimately impacted the community culture, from conferences and collaborations to the publication process.

For the most part, the community divide came down to whether one decided to adopt and adhere to the "criteria of authenticity" as outlined in Cooper and Poinar's article. A younger researcher recalled this particular phase of this discipline's development: "'[Ancient] DNA: Do It Right or Not at All' is like the nucleus of one era of *total proprietary*. 'Ancient DNA belongs only in Ancient DNA Labs. If you don't have this—if you don't do it *exactly* as we say—you're out. We won't publish your research'" (Interviewee 27). One leading practitioner, for example, compared the divide to a "religious schism" where people "stopped talking to each other" and "dissed each other" and "sabotaged each other." This resulted in "different conferences" with "different prophets" (Interviewee 6).

In their attempt to control contamination in light of celebrity concerns, certain scientists were also attempting to control community competition by making it more difficult for some to participate in the practice. According to a molecular evolutionary biologist in the field, for example, "criteria of authenticity" seemed to be, at least to a certain degree, "a means to control or limit access" to search for DNA from fossils (Interviewee 2). Another early researcher in the 1990s explained it this way: "Ancient DNA has really been run by . . . Svante [Pääbo] and [Alan] Cooper. Cooper's role in steering things or pushing things in certain ways, criteria of authenticity, or 'Do It with Me or Not at All School of Ancient DNA' . . . —I find it very annoying because it's not a nice field to work in when everybody is saying, 'You can't!'" For this scientist, this exclusive attitude was a problem, even unscientific: "There is no 'one way' because science doesn't go the way 'one person' says it" (Interviewee 11). As one practitioner observed, "For some time many people thought—and a few people inside the field tried to make people think—that ancient DNA was kind of something magic, and that only two or three labs in the world can do it" (Interviewee 15). A leading geneticist confirmed just as much: "A lot of people know Svante Pääbo and two or three other people and they completely ignore the other ones" (Interviewee 48). For these interviewees, criteria functioned to control success in the field.

Overall, boundary-work was most evident by way of the division between believers and non-believers—a split that manifested itself via the formation of different associations, conferences, and publication strategies. An unpublished paper, initially intended as a chapter of an unpublished book on ancient DNA research, offers an especially illustrative example of the presence and consequence of boundary-work. According to the chapter's authors, Bernd Herrmann at Georg August Universität Göttingen in Germany and Charles Greenblatt of the Hebrew University of Jerusalem, "The scientific community has not acted optimally in establishing a supporti[ve] and cooperative system, but has pretty early started in splitting up in schools of the 'haves' and the 'have-nots' instead." For Herrmann and

Greenblatt, both of whom found themselves on the side of the believers (or have-nots), the non-believers "partially defined themselves more in terms of expert knowledge by self allocation of scientific standards." Specifically, they argued that in the field's contest for credibility, the believers turned "basic epistemological standards of experimental design and falsification" into a "battlefield."[45] In an interview, another practitioner offered a similar sentiment, humorously describing the field as a "feudal system with lords battling each other," "trying to gain control," and "setting forth on horses to destroy each other's kingdoms" (Interviewee 25). Indeed, boundaries, intended for disciplining the discipline, were more than a means to address concerns about contamination and celebrity.

For practitioners, boundaries were also a means of controlling competition, thus shaping ancient DNA's disciplinary development, community culture, and even how scientists approached the writing of its history. In certain cases, practitioners constructed their own history by dismissing other research they disagreed with but that was nonetheless a part of the field's past. Herrmann and Greenblatt, for example, argued that "different positions were not discussed in terms of scientific standards but were ignored rather by strategic behavior" like "citation cartels" and "self referential structures."[46] Indeed, the first textbooks in the field, such as one by Herrmann and Susanne Hummel, cite several different studies as compared to research literature reviews written by Pääbo, Cooper, and students trained in their tradition.[47] Some non-believers—like Pääbo, Cooper, and others—disregarded research by the believers, who included scientists like Herrmann and Hummel. As one senior practitioner explained, "I think we would certainly have a feeling that there is a kind of body of work that we simply don't believe, in the past, and that we don't cite rather than spending a lot of time saying we don't believe it" (Interviewee 36). In an article, Eske Willerslev, a student of Alan Cooper, and Cooper himself also noted the tendency to exclude certain studies, for a number of reasons, from the literature: "Perhaps unsurprisingly, many of the most extravagant a[ncient] DNA reports have since been either disproved or effectively disregarded. . . . Many other claims remain in limbo, where a lack of appropriate methods or replication renders them effectively meaningless."[48] The writing in, or writing out, of certain studies and their authors from the history of the discipline was a further form of boundary-work that some scientists sought in an effort to establish their preferred version of the field as well as their own places within it.

THE ROLE OF AUTHENTICITY

In the wake of contamination problems at the turn of century, researchers felt their credibility was compromised. For some scientists, such as Cooper and Poinar, it was more than the credibility of one or two or even a handful of studies at stake. Rather, they believed the credibility of the discipline as a whole was in jeopardy. The one way to fix this, as far as they were concerned, was for fellow colleagues in the field to strictly adhere to the criteria of authenticity. Here, ancient DNA researchers engaged in boundary-work through the publication of criteria of authenticity and a mandate that fellow colleagues, editors, and reviewers alike adhere to these standards. In some ways, a handful of practitioners advocated for technical boundaries around the practice through the implementation of criteria via the use of certain technologies and techniques. In other ways, this boundary-work was quite literal as the presence or absence of an "Ancient DNA Lab" was a physical boundary used by some scientists to help demarcate credible from less credible work. The criteria of authenticity, and the call to reproduce results, became a hallmark of experimental expertise by which ancient DNA research was judged.

By 2005, however, researchers were realizing that the criteria for ancient DNA authenticity were not infallible. Practitioners, for example, were finding that DNA degradation patterns and processes were not well understood despite technological and conceptual advances. In response to this, Thomas Gilbert and Michael Hofreiter, former students of Cooper and Pääbo, respectively, along with Hans-Jürgen Bandelt and Ian Barnes, challenged the use of criteria that had been introduced in Cooper and Poinar's landmark paper. In this new article, the authors noted that Cooper and Poinar had initially intended the criteria as a *guide* for ancient DNA authenticity. In practice, its use had turned into a *checklist* that some thought actually *guaranteed* authenticity of research results. According to Gilbert and colleagues, what began as a guide became a "religious doctrine" that some "blindly followed." This was particularly a problem because researchers were beginning to realize that these parameters did not always ensure authenticity. In fact, they could, in certain cases, produce false results. Although the criteria were useful, these authors argued that treating criteria as a checklist had become an inappropriate substitute for critical thinking and scientific reasoning: "Nevertheless, these criteria are not foolproof, and we believe that they have, in practice, replaced the use of thought and prudence when designing and executing ancient DNA studies."[49]

Such dogmatic adherence to the checklist criteria had produced two problems. In an interview, one scientist explained the issue, saying that in many cases, papers had been published by scientists who had fulfilled all the criteria but despite this, they were still "publishing bad results." At the same time, papers were not being published because scientists had not fulfilled all the criteria, and in some of these instances they were actually "failing to publish good results" (Interviewee 6). This concern was mirrored in Gilbert and co-authors' argument that completion of criteria should not, and in fact could not, serve as the be-all and end-all of ancient DNA research. According to the authors, the "authenticity and reliability of ancient DNA data arise from a complex interplay of several poorly understood areas of knowledge," and "no clear-cut answer exists as to what makes a study reliable."[50] Even after years of debate and disciplinary development, there seemed to be no definite community agreement on what really makes a "good" or "bad" ancient DNA study.

This counterclaim was not necessarily a rejection of criteria for ancient DNA authenticity. Rather, these researchers were bringing attention to the fact that the criteria were imperfect and that rigid dedication to them was problematic. Gilbert and co-authors suggested a solution: "It is our opinion that ancient DNA researchers should take a more cognitive approach with regards to assessing the reliability and conclusions of their data. Suggested criteria remain important, and should not be lightly discarded, but we advocate that, in place of planning or assessing studies by using criteria as checklists, consideration should be given on a case-by-case basis as to whether the evidence presented is strong enough to satisfy authenticity given the problems."[51] The authors argued that scientists should assess their projects by asking questions about feasibility. For example, does the age and environment of the sample suggest DNA preservation? Or is there information about the handling history of the sample that might suggest prior contamination which might be difficult to detect and therefore jeopardize ancient DNA authenticity?

Interestingly, Pääbo and Hendrik Poinar, as well as Cooper and Willerslev, had made similar suggestions in the years following the publication of "Ancient DNA: Do It Right or Not at All" as they too began to realize the inherent issues with their criteria.[52] Reflecting on the article's publication, an influential scientist in the field said, "And the *big* mistake . . . with that paper—*huge* mistake . . . with that paper—was not putting at the end of that list: '*If* the result passes the criteria, it's probably still wrong. It's just that you failed to disprove it'" (Interviewee 32). Nonetheless, these criteria and the severity with which they had been enforced continued to provoke further debate in the community.

Although controversy over what counts as a thorough experiment is common across the sciences, this is a particular problem for new fields of science or in areas of controversial science. "The problem with experiments," Harry Collins and Trevor Pinch write, "is that they tell you nothing unless they are competently done, but in controversial science no-one can agree on a criterion of competence. . . . Thus, in controversies, it is invariability the case that scientists disagree not only about results, but about the quality of each other's work. This is what stops experiments being decisive and gives rise to the regress."[53] Ideally, "reproducibility" should serve as evidence for the "universality" of science: "Anybody, irrespective of who or what they are, in principle ought to be able to check for themselves through their own experiments that a scientific claim is valid."[54] However, in the case of an emerging science or an area of controversy within a science, the "who" and "what" and "how" of an experiment become subjects of debate. The intensity of this debate around contamination was compounded by the celebrity that surrounded the science.

As researchers engaged in boundary-work, they did so on two fronts and in response to two separate but intertwined issues impacting their legitimacy with the broader public and their authority in the scientific community.[55] While they were concerned about contamination in the literal and technical understanding of the term, they were also concerned about the influence of celebrity. On one hand, some researchers working both in and outside of the discipline viewed the ever present media attention around the search for DNA from fossils as a further and more figurative source of contamination but one that was no less real. They felt that disproportionate and undeserved media interest or influence had the adverse effect of contaminating their credibility, thus their scientific authority. On the other hand, media attention had been a crucial component of the field's growth in terms of its initial formation and overall identity. Over the years, the press consistently publicized the nascent science. Meanwhile, scientists also fashioned their own opportunities for exposure. This intentional exchange between scientists and the media—specifically around the idea of discovering DNA from some of the world's most ancient and charismatic creatures, such as dinosaurs or mammoths—influenced scientific practices from research agendas and student recruitment to publications and further funding. In fact, the public's interest in the search for DNA from fossils and the attention the media gave to it were instrumental in its birth and growth into a scientific discipline.

Consequently, researchers were torn between their need to appeal to the press and public for support while simultaneously distancing themselves from the hype that had come to characterize the field. In speaking of science in general, cultural studies scholar Peter Broks summarizes the dilemma: "To maintain its authority it needs to be set apart from the general public, but to maintain its legitimacy it needs to appeal to the general public." However, in practice, this presents an unavoidable tension: "Being set apart increases its alienation; making it more 'popular' undermines its authority."[56] Indeed, ancient DNA researchers found themselves walking a fine line. It was difficult for researchers to balance legitimacy and authority, especially in light of increasingly apparent technological limitations and contamination issues. At the turn of the twenty-first century, members of the ancient DNA community found themselves facing a paradoxical situation as the celebrity of the science simultaneously empowered and undermined it.

CHAPTER SEVEN

Ancient Genetics to Ancient Genomics

NEXT-GENERATION SEQUENCING

In 2005, 454 Life Sciences Corporation—a biotechnology company in Branford, Connecticut—announced the innovation of next-generation sequencing (NGS).[1] Created in part by Jonathan Rothberg, the founder of the company, along with several dozen researchers, NGS was the result of years of work dedicated to developing a more time-efficient and cost-effective DNA sequencing technology. The rise of large-scale sequencing projects, including whole-genome sequencing, and the need for a machine that could automate the workload motivated their pursuit. The Human Genome Project, which was initiated and completed just before the advent of NGS, offers a case in point. That project, begun in 1990 and finished in 2003, was an unprecedented international endeavor involving thousands of scientists, over a decade of research, and just under $3 billion to determine the order of nearly 3 billion base pairs of DNA that together make up the human genome.

In contrast, NGS made whole genome sequencing much easier by dramatically increasing the speed of production while decreasing the total cost. Indeed, the beauty of this state-of-the-art technology was in its unmatched throughput. NGS could produce close to one billion sequences in a single run over the course of just a few days. With this technology, it was possible to efficiently sequence an entire genome, a process that includes identifying all the DNA that makes up an organism and determining the exact order of all the bases or letters (A, G, T, and C) in each strand of DNA. In 2007, for example, scientists sequenced the genome of James Watson—one of the co-discoverers of the helical structure of DNA—in a mere two months and for less than $1 million, "a 1,000-fold improvement over the cost of the decade-long Human Genome Project."[2] Practitioners have called NGS nothing less than a "paradigm shift" into a new era of technological innovation and scientific possibility.[3]

NGS has been used as a broad term to characterize a number of high-throughput sequencing technologies, a variety of machines that use parallel platforms to sequence more than 1 million short reads of DNA (50–400 base pairs) at one time. The technology has had a profound impact on the field of genomics more generally, revolutionizing both the scale and scope of research that is possible. Although NGS was not developed to aid the search for DNA from fossils, researchers quickly recognized its benefits and potential to transform their own work. There were several platforms available (varying in chemistry and technology), but two instruments in particular became widely used in ancient DNA research during this period: Roche (454) GS FLX and Illumina (Solexa) Genome Analyzer.[4] Because ancient DNA is often damaged and fragmented, researchers have been constrained by short DNA sequences. However, NGS favors short sequences. Thus, what was once a disadvantage for ancient DNA research has now become an advantage. Overall, the technology has enabled practitioners to generate a much higher quantity and quality of data in a fraction of the time and cost when compared to previous sequencing approaches.

Most obviously, the difference between the pre-NGS and post-NGS era was the ability to generate a handful of sequences as opposed to billions of sequences. Previously, the study of ancient DNA, because of its degraded and damaged state, had been limited to the study of mitochondrial DNA and occasionally nuclear DNA. Mitochondrial DNA is the most accessible form because of its abundance in animal and plant cells, which means there is a higher likelihood that at least some genetic material would be preserved and could then be extracted. Mitochondrial DNA, inherited from the maternal line, is informative but offers only a partial picture of an organism's genetic history. However, mitochondrial DNA in combination with nuclear DNA, inherited on the paternal line, offers a more complete picture. With NGS, it became possible to easily sequence any and all DNA from a sample, thus making it theoretically possible to recover the entire genome of long-dead creatures at nearly the same rate, as well as cost, of that of living species.

Two separate studies—issued just six months apart—exemplified the drastic impact that NGS could have on the field of ancient DNA research in particular. In the first study, published in 2005 by *Science*, researchers recovered nearly 27,000 base pairs of ancient genomic data from two 40,000-year-old cave bears. The study—by James P. Noonan and Eddy Rubin (both at the U.S. Department of Energy Joint Genome Institute and Lawrence Berkeley National Laboratory in California) and conducted in collaboration with colleagues including Michael Hofreiter and Svante Pääbo at the MPIEVA in Leipzig—used a direct-cloning technique that did not involve the traditional PCR amplification of targeted sequences as had previously been common in the field of ancient DNA research. The technique they developed was different and intended to circumvent issues associated with PCR while generating a higher quantity and quality of DNA from their specimen of study. With this technique, they were able to sequence more than just the genome of the ancient cave bears. They were able to sequence its metagenome, meaning the collective genomic sequences of all the organisms associated with a single sample. Here, the metagenome was a mixture of the ancient DNA sequences of the organism of interest, in this instance the cave bear, plus any DNA sequences from other organisms and the external environment that had come in contact with the specimen. At the time of publication in 2005, their results represented the largest data set of ancient DNA sequences from an extinct species, and the study overall showcased the potential to access the entire genome of ancient organisms.[5]

This feat, however, was overshadowed less than a year later. In a second study, also published by *Science*, Hendrik Poinar—recently appointed head of the Ancient DNA Centre at McMaster University in Ontario, Canada—and colleagues capitalized on the recent availability and advantages of NGS to successfully sequence 13 million base pairs of ancient genomic data from a 28,000-year-old woolly mammoth. To generate this amount of data, the team used a technique referred to as shotgun sequencing. In the pre-NGS era, PCR and Sanger sequencing were used to target a specific DNA sequence. In contrast, NGS (in combination with shotgun sequencing) made it possible to sequence all the available DNA in a sample. In this process, DNA is randomly broken into numerous short overlapping strands, then cloned, sequenced, and finally reassembled. In this case, when the sequences were reassembled, scientists found that the mammoth remains contained much more than just mammoth DNA. Scientists sequenced a total of 28 million base pairs of DNA of which 13 million base pairs were identified as authentically ancient mammoth DNA. The remaining data, approximately 15 million base pairs, was environmental, bacterial, or unidentified DNA.[6] The difference between the two papers in terms of data output was clear. The recovery of 13 million base pairs of genomic data from the woolly mammoth, compared with the approximately 27,000 base pairs of genomic data from the extinct cave bear, was an impressive 480 times increase in yield.[7] "The change was massive," said one scientist, "absolutely massive" (Interviewee 15).

In a review paper, Alan Cooper—formerly at Oxford and at the time recently appointed to the Australian Centre for Ancient DNA at the University of Adelaide—emphasized how NGS could affect the field. To make his point, Cooper spotlighted three near back-to-back articles, published within six weeks of each other, all claiming the independent and near complete sequencing of the mammoth genome. For Cooper, this series of studies conveniently captured what he saw as the past, present, and future of ancient DNA research in regard to its technological potential.[8] Although each study accomplished similar achievements, they did so through distinctly different techniques. The first paper, by geneticist Evgeny I. Rogaev and colleagues, used PCR, while the second paper, by Johannes Krause and co-authors, used a multiplexing method, a variation of PCR that simultaneously amplifies multiple targets, as opposed to just one.[9] The third was, of course, Poinar and colleagues' landmark paper reporting the recovery of the mammoth metagenome—nearly 28 million base pairs of genomic sequences—in just one experiment by using the new high-throughput sequencing technology developed by 454 Life Sciences.[10] For Cooper, the first two represented the past and present state of the field, respectively, while the third provided the possibility for a technical transition into the future. Collectively, these papers and the diverse methods employed provided a conceptual and technological snapshot of the discipline's history and conceivable future. "This is an exciting time," wrote Cooper, "as the opportunities by the new parallel sequencing system will allow researchers to contemplate large-scale studies of ancient genomes, and promise to finally release the full potential of a[ncient] DNA to reveal evolution in action."[11] Given the potential, a number of researchers in the ancient DNA community eagerly embraced the new technology.

For this community specifically, NGS offered an opportunity to over-come some of their most persistent technological challenges, namely the low quantity of data and the problem of contamination. To be clear, NGS did not remove the possibility of contamination, but it did reframe the problem. Ancient samples indeed had contaminating sequences, but practitioners were able to calculate the amount of contamination, permitting them to increase confidence in DNA authenticity. They did this by reading the sequence data and looking for molecular signatures of chemical degrada-tion, usually postmortem damage (changes that occurred in the organism after death) that were characteristic of authentically ancient DNA. Further sophisticated computational techniques were also needed to estimate the amount of contamination. The availability of more data thanks to NGS, combined with scientists' ability to recognize and analyze patterns indicative of DNA degradation, made it possible for researchers to determine which sequences were from the organism itself and which belonged to the external environment. Although lab and specimen handling protocols remained stringent (in order to prevent further unnecessary contamination), ancient DNA researchers could, in a sense, rest in the fact that they could actually estimate amounts of contamination. "So, now it's not only a question of having controls," explained an interviewee. "You can actually look at your data and determine whether you have a contamination problem or not, right?" (Interviewee 7). According to a second scientist, the somewhat newfound freedom from contamination concerns was of great consequence for the research community: "I remember *fighting* at conferences [about] if these sequences [were] feasible or not; if it was contamination or not." But with NGS, the focus and fighting over results had changed: "This is not really an issue anymore because people have contamination, but they calcu-late it away. [Laughs]" (Interviewee 13).

The newfound ability to more easily differentiate between authenti-cally ancient DNA and contemporary contaminating DNA was a major advantage of NGS. Indeed, the community's near decade-long debate over the cause of the Black Death, as evidenced through ancient genetic data, was one of the most salient examples of how NGS would change the discipline in terms of contamination concerns. The Black Death—one of the most destructive pandemics in human history—killed millions of people across Europe in the mid-fourteenth century. Although there was much historical speculation about its biological cause, definitive evidence had yet to be found. In 2000, a team of scientists—including Didier Raoult and Michel Drancourt from the University of the Mediterranean in Marseille, France—tried to answer the question with ancient DNA. The team sampled human remains from a mass burial site in the south of France. From these remains, they recovered DNA sequences for a particular bacterium, *Yersinia pestis*. According to their paper, published in the *Proceedings of the National Academy of Sciences,* they had found the cause of the Black Death, ultimately solving a six-hundred-year-old mystery.[12]

At this time, however, five years before the invention of NGS, the field was still at the height of controversy regarding contamination. It was not long until another team challenged Raoult and colleagues' conclusions. A group led by Thomas Gilbert—a postdoctoral researcher working with Cooper, who was then at Oxford—attempted to extract and identify the bacteria from more than a hundred samples taken from mass burial pits across Europe dating to the plague. Despite their seemingly comprehensive sampling, they failed to replicate positive results.[13] One researcher recalled the debate that ensued: "So, you had Didier saying, 'We found it!' And then Tom would say, 'You didn't find it!' 'We found it!' 'You didn't find it!' And there's probably ten years of publications going back and forth about this" (Interviewee 27).

Several years later, another group—independent of either side of the debate—gathered its own samples to be sequenced using the newfound availability of NGS.[14] Kirsten Bos, a doctoral student supervised by Hendrik Poinar in Ontario, collaborated with colleagues to sample DNA from teeth found in a plague pit in central London. In combination with NGS, they applied a specialty technique—targeted capture—that allowed them to pinpoint sequences of interest while leaving behind other genetic material. Krause, another author on the publication and now a professor at the University of Tübingen, had learned this technique while a graduate student in Pääbo's lab in Leipzig. Not only did they obtain evidence for the bacteria, *Yersinia pestis,* but they were able to sequence the genome. In doing so, their study settled the dispute: *Yersinia pestis* was the cause of the Black Death. One researcher remembered the mark this case left on the field: "Once they had the whole genome then there was no question, right? It completely ended the debate. . . . It dropped like a *bomb* on the community—like a *huge bomb*. [Laughs]" For scientists at the time, it was a pivotal moment. "It was also one of the earliest demonstrations of [how] next-generation sequencing is going to completely change this game," recalled this researcher. "We're in a different era, and it just shut that whole thing down" (Interviewee 27).

However, in the years between Cooper and Poinar's article in 2000 and the innovation of NGS in 2005, ancient DNA researchers were not waiting idly for the next best technology to come along. A handful of practitioners were persistent in their search for DNA from fossils, seeking—despite the odds—to find more efficient approaches to recovering greater quantities of and better quality genetic material from ancient specimens.[15] Researchers realized that a handful of sequences from one or two specimens was of limited value. They needed more fossils, more DNA. They needed confidence of authenticity too. This shift signified a growing awareness among researchers that if the search for DNA from fossils was to become relevant to the evolutionary biology community, they would need to answer questions on the population level, not just the individual level. In an effort to achieve this, some scientists set out to recover and analyze DNA from a wide range of ancient samples.

¡   One of the earliest examples exhibiting a move in this direction was a paper by Jennifer A. Leonard, Robert K. Wayne, and Alan Cooper in which they successfully obtained sequences from seven permafrost-preserved brown bear specimens dating back to the Ice Age.[16] With this data, they demonstrated that the present distribution of brown bears in terms of demography and geography was distinctly genetically different from its past. Ancient DNA showed a side of the story that was indiscernible with inferences from modern genetic material alone. One scholar observed, "The big change in terms of moving into population genetics—away from phylogenetics—[was] to provide population genetics with a time scale which it had never had before." According to this interviewee, the paper presented a "big conceptual breakthrough" (Interviewee 32). Another early example of this transition was a study led by Beth Shapiro, a doctoral researcher with Cooper at Oxford.[17] This study was conceptually and technologically important for its large number of samples, use of statistical demographic modeling, and conclusions that contested previous assumptions about bison evolution and extinction. Overall, this study—among others similar to it—demonstrated the potential to obtain larger amounts of ancient genetic data in order to test hypotheses about the evolution and extinction of past populations, as well as its impact on conservation biology and understandings of climate change.[18]

NGS was in part so important to the field because it presented practitioners with a chance to much more easily and efficiently transform the search for DNA from fossils into a credible practice capable of addressing bigger questions within evolutionary biology. Given its advantages, NGS surpassed Sanger sequencing, the primary sequencing technology since the late 1970s, and it soon overshadowed PCR, the once state-of-the-art technique that sparked the field of ancient DNA research into existence but now held it back in light of technical limitations. "I used to joke that I was a retired ancient DNA researcher," said one scientist, "but then the big game changer— without a shadow of a doubt—has been ultra-high-throughput sequencing or next-generation sequencing. And it has completely rescued the field." Comparing the eras of PCR versus NGS, this same scientist emphasized the variance in data output between the technological paradigms: "PCR allowed ancient genes, NGS has allowed ancient genomes" (Interviewee 21).

NEANDERTHAL GENOME

In July 2006—shortly after the introduction of NGS and its initial application to a number of ancient DNA studies—Pääbo with the MPIEVA, along with 454 Life Sciences, announced they would be the first to attempt to sequence the entire Neanderthal genome. They planned to sequence the first genome of our extinct and archaic ancestor, and they would do it in just two years' time. All of this was announced for the first time through a highly orchestrated press conference and press release.[19] In his memoir, Pääbo remembered the press conference as an "electrifying event"; the "room was full of journalists," and "media from across the globe" tuned in online.[20] It seemed that everyone wanted to hear what scientists and the public had once thought impossible. From the outset, the Neanderthal Genome Project was a substantial media production. Furthermore, what was broadcast as a big event was going to be an even bigger effort. The fact that an accomplished yet careful and conservative practitioner such as Pääbo would both initiate and publicly advertise a venture of this magnitude was evidence of his confidence in NGS to help them deliver this extraordinary achievement.

The thrill of such an unprecedented endeavor was followed by the stress of work to be done in a short time span. Pääbo was acutely aware of the pressure he had placed on himself and his lab. "Now I had really stuck my neck out," he noted in his memoir, "publicly promising to sequence the Neanderthal genome." Indeed, the stakes were high: "If we succeeded, it would clearly be my biggest achievement to date; but if we failed, it would be a very public embarrassment, almost surely a career-ending one." He admitted that succeeding would not be as easy as he made it sound. Indeed, a mere two months before the project's official announcement, Pääbo had presented his plans to fellow scientists at the Cold Spring Harbor Laboratory's Annual Symposium on Genome Biology. At that time, he and colleagues had just sequenced nearly 1 million base pairs of Neanderthal DNA. But a million base pairs, although a real feat, was far from what they needed to eventually reconstruct the entire genome, which would include almost 3 billion base pairs. The present data represented only 0.0003 percent of the whole genome.[21] Nonetheless, he claimed that it could, and would, be done. As far as Pääbo was concerned, in principle it was possible.

Pääbo's colleagues and collaborators felt the gravity of the situation too. Not only was the project a technical and financial challenge but the media attention magnified the pressure to perform and perform well. "The pressure we had," explained one interviewee involved with the project, "was a self-inflicted pressure that Svante had created by announcing that we would publish the genome in two . . . years or something crazy." Pääbo's lab was a powerhouse institution in the field, but even with its technical expertise and financial access, it was underequipped to reach the goal. "We didn't even have the material to do it," recalled the same scientist. When the project was proposed, its attainment really rested on the idea that there would be technical and methodological improvements, hopefully sooner than later (Interviewee 12). They needed more money, more machines, well-developed techniques, and, most importantly, well-preserved fossils with Neanderthal DNA.

The Neanderthal Genome Project, although a unique effort on its own, was not an isolated idea. Rather, it was the product of major technological advances coupled with widespread interest in whole genome sequencing projects.[22] The Human Genome Project, for example, was a herculean effort requiring an exceptional amount of talent, money, technology, and resources. Leading up to its launch and throughout its duration, the project was advertised by scientists, reporters, and politicians alike as a "Holy Grail" for understanding life itself.[23] Additionally, the Neanderthal Genome Project was also the product of various scientific, conceptual, and technical developments seeking to study the evolution and extinction of Neanderthals through DNA.[24] After Pääbo's lab at the University of Munich and Mark Stoneking's lab at Pennsylvania State University were the first to successfully sequence Neanderthal DNA, they explained how they found no evidence that Neanderthals and our ancient human ancestors had interbred thousands of years ago—but they also noted that this conclusion could not be definitively determined by mitochondrial DNA alone; they needed genomic data and lots of it.[25]

NGS offered an opportunity to sequence a higher quantity of Neanderthal DNA with the possibility of yielding better quality data too at a much lower cost in terms of time and fossil material. Initially, Pääbo and Eddy Rubin—a biophysicist turned geneticist at Berkeley—agreed to collaborate since they had recently worked together on the sequencing of several thousand base pairs of ancient genomic data from forty-thousand-year-old fossil cave bear remains. For this collaboration, however, they would be working on a fossil species that was much rarer. The MPIEVA sent Berkeley an extract from a 38,000-year-old Neanderthal fossil that came from the Vindija Cave in northern Croatia with the intention that their respective labs would attempt to sequence authentic Neanderthal DNA. From the outset, however, Pääbo and Rubin disagreed on exactly how they would go about extracting and sequencing the DNA. Rubin was set on indirect sequencing via recent advances in traditional bacterial cloning methods, while Pääbo insisted on direct sequencing via NGS, arguing that this approach would permit them to sequence more DNA using less fossil material. At first, they agreed to disagree. Rubin's lab employed its indirect sequencing approach, recovering 36,000 base pairs of Neanderthal DNA. Meanwhile, Pääbo's lab used its direct sequencing approach with NGS and recovered nearly 750,000 base pairs of Neanderthal DNA.[26] Neither lab had yet sequenced the whole genome, but they succeeded in recovering a partial draft and planned to publish the results in anticipation of more to come.

Soon, however, Rubin's and Pääbo's differences in their approaches turned to discord. Not only had Rubin and Pääbo implemented very different methods but their respective methods resulted in a sizeable difference in the amount of data obtained. It became clear to both that they would have to publish separately.[27] On November 16, 2006, *Nature* published the MPIEVA's research led by Richard E. Green—a bioinformatician and recent postdoctoral researcher in Pääbo's lab. The next day, *Science* published Berkeley's findings, led by James P. Noonan in Rubin's lab. As back-to-back publications in world-renowned journals, the disparities between the papers were obvious. Not only did each group use different methods but they also arrived at clearly different conclusions. Data from Rubin's lab provided no evidence for the genetic contribution of Neanderthal DNA to modern humans. Conversely, results from Pääbo's lab suggested a significant amount of admixture between the two. "The conclusions of the studies are pretty much completely opposite," recalled a researcher. "One of them says there's no mixing with modern humans, one says there's a lot of mixing with modern humans. And the weird thing is they both analyzed the same bone. So, it wasn't even two different Neanderthals" (Interviewee 6). Soon after publication, a team of different researchers—independent of either lab—reanalyzed both data sets in light of their contrasting conclusions. In the end, they found evidence of contamination in the Neanderthal DNA sequenced in Pääbo's lab. Specifically, they found evidence of modern human DNA, which explained why that lab's conclusions supported admixture between Neanderthal and modern humans, thus conflicting with the data and conclusions from Rubin's lab.[28]

According to Pääbo's memoir, he and his lab in Leipzig had worried about contamination in their own findings, so much that they considered rewriting, even retracting, their paper awaiting publication in *Nature*. To be sure of their findings, they sent their data to Rubin's lab for comparison. Indeed, Rubin's lab confirmed that Pääbo's lab had a level of contamination in their results based on differences in sequences. As far as Pääbo was concerned, the differences could be from bacterial contamination or even the result of genetic mutations. In response, Pääbo's lab frantically sequenced and analyzed the results again and was able to measure the likelihood of contamination by comparing a set of Neanderthal DNA sequences with those of modern humans. Based on the fragments they reanalyzed, they determined that they indeed had recovered authentic sequences from the Neanderthal specimen under study and that the level of contamination was low. Accordingly, they reasoned that the differences in sequences their lab, as well as Rubin's lab, had seen were perhaps the result of other unknown factors and not direct evidence of contamination. Thus, Pääbo's lab decided to publish anyway.[29] They would get these new results out and analyze the anomalies later.[30]

The community of ancient DNA researchers heavily debated the publications and their differing conclusions. Not only did one of the publications appear to suffer from problems of contamination but the lead on this study was Svante Pääbo, a symbol of conservatism regarding the risk of contamination in ancient DNA studies. Over the years, he had made a name for himself and his lab against contamination. Now it appeared that he and his lab had published results with knowledge of contamination, or at least knowledge of possible contamination. "There are rumors even that he submitted it knowing it was contaminated," recalled one researcher. "And one of the supporting statements for that is the fact that Eddy Rubin is not on Svante's paper but that Svante is on Eddy's paper, suggesting that Eddy withdrew himself from Svante's paper because he knew there was something wrong with it. . . . But the interesting thing is that Svante never published an errata on that and it's kind of weird given it's the standard behavior" (Interviewee 6).

Pääbo himself noted the tension, especially in his collaboration with Rubin. Pääbo and Rubin had previously disagreed on the approaches they would use to sequence the Neanderthal genome, and after they published their papers, it became quite clear (at least to Pääbo) that if Rubin was not going to collaborate with him toward the Neanderthal genome, he would compete for it. According to Pääbo, Rubin was after the same Neanderthal bones, from the same individuals or institutions that they had both worked with together for years.[31] In his memoir, Pääbo quoted Rubin in an interview with a media reporter from *Wired,* which had Rubin saying: "I need to get more bone. . . . I'll go to Russia with a pillowcase and an envelope full of euros and meet with guys who have big shoulder pads. Whatever it takes."[32] Motivated by fear that Rubin would publish the entire genome first, Pääbo impressed on his lab the need to complete the project as soon as possible.[33]

In 2010, a decade after the first Neanderthal sequences were recovered and four years following the initial announcement of the Neanderthal Genome Project, the MPIEVA at long last published a first complete draft of the Neanderthal genome.[34] They were first to the finish line. The project, conducted by over fifty scientists at a cost of approximately 5 million Euros, successfully sequenced more than 4 billion base pairs of Neanderthal DNA obtained from three different individuals.[35] What made the project and the paper so impressive, however, was the scientists' analysis of the data and their interpretation of its implications for understanding human evolution. Pääbo enlisted David Reich, a population geneticist from Harvard University, to help make sense of all the data. Indeed, Reich played a lead role in the project's overall success. Data alone would not be enough; the tools for its analysis were a critical component of the project. It was the combination of genomic data generated by Pääbo's lab and statistical methods developed by Reich's lab that allowed them to detect signals of admixture between humans and Neanderthals. In other words, the data analysis suggested clear and extensive evidence that early humans had interbred with their archaic ancestors the Neanderthals before they went extinct nearly forty thousand years ago.

Crucially, the evidence for admixture seemed to suggest that Neanderthals only interbred with a particular human population, those early peoples who had traveled out of Africa into Europe. By comparing the Neanderthal genome with modern human genomes across the world, scientists determined that Neanderthals shared more similarities with present-day non-African populations than with present-day African populations. Neanderthal DNA existed in a small percentage (1–4%) of a specific population (Eurasian population). In other words, humans today of European or Asian descent, but not African descent, have bits of Neanderthal DNA in their own DNA. "The next time you're tempted to call someone a Neanderthal," reported *National Geographic,* "you might want to take a look in the mirror."[36] Sure enough, some humans may have more in common with our extinct Neanderthal cousins than previously imagined.

Pääbo expected the Neanderthal Genome Project and the announcement of its findings to have a significant impact on the archeological and anthropological communities, but he claimed, however naively, not to have anticipated the public reaction. Indeed, the surprising conclusion about interbreeding with Neanderthals generated a massive amount of media attention. According to Pääbo's memoir, his paper published in *Science,* for example, attracted attention from the creationist community, a conservative fundamentalist religious group in the United States, who reinterpreted the results as evidence in favor of their own views about Neanderthals' relations to humans and creation.[37] Several random women wrote to Pääbo speculating that their own husbands were in fact living, breathing Neanderthals in the modern age. *Playboy* even spotlighted the research in a four-page spread titled, "Neanderthal Love: Would You Sleep with This Woman?"[38] These reactions were hardly surprising given the enduring public interest in the science of ancient DNA research and the study of human evolution. The Neanderthal Genome Project was packaged, pitched, and even intentionally pursued within this context, and with an awareness of its scientific significance, as well as its news value.

GENOME REVOLUTION

The search for ancient genomes, facilitated by the technical convenience of NGS and its ability to produce genomic data quickly and relatively cheaply, has ushered in a race among researchers to be the first to sequence whole genomes from a variety of specimens including ancient plants, animals, and diseases. Some researchers have also set their sights on recovering genomic information from ancient humans including Paleo-Eskimos, Aboriginal Australians, and famous historical figures like King Richard III.[39] In the search for ancient human genomes, scientists have used this data to shed light on the behavior of our early ancestors, including Mesolithic and Neolithic hunter-gatherers, while also exploring transformations in human cultural practices such as milk consumption, which have directly impacted our evolution in terms of selection for lactase persistence.[40] Much work has examined our interactions with animals through time by interrogating genetic signals for domestication in pigs, cattle, and dogs on large global and temporal scales.[41] Using NGS, scientists are seeking to reach farther back in time to learn more about our extinct and archaic ancestor the Neanderthal. With this genome-wide data, scientists were able to estimate the extent to which early humans and Neanderthals had interbred before the latter's extinction.[42] Adding to the excitement, practitioners sequenced the first genomic data from a Denisovan, a formerly unknown extinct hominin species whose identity as a distinct archaic human species was uniquely obtained from DNA extracted from a small finger bone, as no extensive fossil record exists.[43] According to scientists and journalists, these works— among others—represent a revolution in our understanding of human history in terms of our origin, evolution, and migration across the globe.[44]

For some scientists, this adaptation of high-throughput sequencing technologies to the search for DNA from fossils, and the massive amount of data that could be produced from it, as well as the resulting grandiose conclusions that researchers could draw from it, suggested an overcoming of previous limitations and a maturation of the field. At the same time, however, the field seemed to be coming full circle, back to an era of exploration and hype. This race for the first or the oldest genome (as well as the race to sequence the most genomes), and the accompanying media attention surrounding these high-profile publications in *Nature* and *Science,* shared striking similarities to the search in the 1990s for the first or oldest DNA. In one way, this hype took form through scientists' newfound confidence not only in the technology of NGS to generate a higher quality and quantity of genomic data from ancient specimens but also in their ability to overcome previous contamination concerns. In another way, hype also took form through scientists'—as well as media reporters'—projections that the field had come of age and into a new role as an authority on human evolutionary history as told through DNA. As ancient DNA researchers explored the potential afforded by next-generation sequencing technologies, these two kinds of hype colored the direction of the discipline and the public's perception of it.

Although the introduction of NGS to ancient DNA research did not wholly sweep away the criteria of authenticity as it related to PCR and the decades of debate around the issue, its incorporation into the field did bring a fundamental restructuring of the practice regarding the research questions scientists could ask and the types of resources needed to answer them. There were three specific ways in which NGS altered the nature of the search for DNA from fossils. First, its obvious utility, and scientists' ability to adapt it to their unique pursuit for DNA from fossils, changed the field in terms of scale and scope of data production.[45] High-throughput sequencing technologies could produce an astounding amount of genomic information that required both large amounts of data storage and newfound skills in data analysis. "Processing is completely different because before I could still look at each sequence by eye and edit them by hand, but now we have . . . billions of sequences and you have to do everything by bioinformatics," said an evolutionary biologist. "So, that has changed completely" (Interviewee 15).

Second, this massive increase in data required researchers to learn or seek specialized mathematical, statistical, and computational skills in order to analyze the data and answer questions about evolutionary history. According to another evolutionary biologist, "It's the people who are going to analyze it all that are going to end up with all the work and all the fame and fortune" (Interviewee 25).

Finally, all of this changed the conversation around contamination. Indeed, this shift from ancient genetics to ancient genomics moved the debate from one about data contamination to a focus on data production: "At the moment, we are not discussing the authenticity of the results much anymore," confided one paleogeneticist. "At the moment, we are rather discussing the correct filters that you have to apply to your data set and how to handle these huge amounts of data" (Interviewee 13). According to practitioners, the increased ability to sequence genomes rapidly superseded their aptitude to analyze the data. It even sometimes superseded the questions that they could ask of the data. One interviewee, for example, observed, "People are going *over* the top because they can—just sequencing the living crap out of absolutely everything. So, we're in this kind of exploration phase again, where it's like, 'Grab as much data as you possibly can, hire a great bioinformaticist, and then start asking questions in the resulting data sets'" (Interviewee 22). Together, these changes suggested that scientists found themselves facing a new phase of exploration.

However, the transition from the PCR to the NGS era was not easy. Doing so required both extensive expertise in genetics and bioinformatics and substantial financial resources for sequencing equipment. Even scientists who made the move felt the difficulty in doing so. "It took us a few years, and we're a genetics department," an interviewee explained. "Whereas if someone is in an anthropology or archaeology department, it's quite a different story. It's become, I think, impossible for somebody to transfer from an archaeology or anthropology discipline to this field" (Interviewee 21). As a result, many labs were left behind while others forged ahead. "The kits are expensive, the primers are expensive, and it's all very new," remarked an archeogeneticist. "It was really scary to a lot of labs, and a lot of labs haven't made that transition because it's expensive and it involves the development of a completely new tool set" (Interviewee 27). A lab's decision to transition to NGS-based methodologies was a serious commitment because it was a big intellectual and financial risk.

Nonetheless, some labs made the move successfully, and a select handful made it to the top. Svante Pääbo's lab at the MPIEVA in Leipzig has been one of them. Eske Willerslev, a former postdoctoral researcher with Alan Cooper, at the University of Copenhagen in Denmark is another. In Copenhagen, research has been further bolstered through the work of two intensely productive labs, led respectively by Thomas Gilbert and Ludovic Orlando. Together with Willerslev's lab they make up the Center for GeoGenetics. Indeed, the center's research output, coupled with Willerslev's media-savvy personality, have made these labs internationally famous. Additionally, David Reich—geneticist and collaborator with Pääbo on the Neanderthal Genome Project—developed his own ancient DNA lab at Harvard University in Cambridge, Massachusetts. Although a much more recent recruit, he quickly became a powerful researcher, even competitor, in the field. "Some labs have struck way ahead," one interviewee said. "You know who they are. They're Leipzig, Copenhagen, and Harvard. They're the big productive labs" (Interviewee 21). More recently, Johannes Krause—a former doctoral student of Pääbo's—has emerged as a researcher at the forefront of the field from his recently appointed position as director of archaeogenetics at the Max Planck Institute for the Science of Human History in Jena, Germany.

A shared feature among these labs was the serious financial and institutional support they enjoyed, which enabled them to attract international talent and deliver large-scale, high-impact research. As a consequence, these labs also enjoyed good rapport with leading scientific journals, from *Science* and *Nature* to *Cell* and *Proceedings of the National Academy of Sciences of the United States of America* (*PNAS*), which in turn brought more prestige, along with further access to money and fossil samples. In the process, the heads of these labs became well-accustomed to the media spotlight, having been extensively interviewed and profiled by global media outlets and having established their labs as scientific powerhouses in the field of ancient DNA research, particularly in the study of human evolutionary history.[46] More than that, the work coming out of these labs has claimed to entirely rewrite our understanding of human history. A *New York Times Magazine* article specifically described Pääbo's, Reich's, and Krause's collective influence on the field of ancient DNA research as a "state-of-the-art oligopoly."[47] Indeed, over the past ten years, some of the biggest and boldest claims in the field of ancient DNA research and human evolution have come from this handful of practitioners.

In 2018, Reich published a fairly comprehensive, albeit contentious, book on the evolution of ancient and modern human populations as told by cutting-edge genome-wide data from the field of ancient DNA research. In this book—*Who We Are and How We Got Here*—Reich offers a personal account of his own professional research, as well as that of colleagues, that argues for the power of genetic evidence to tell a new and better story about human history. "Ancient DNA and the genome revolution," Reich claims, "can now answer a previously unresolved question about the deep past: the question of *what happened*—how ancient peoples related to each other and how migrations contributed to the changes evident in the archeological record." He suggests that archeologists should be equally excited by this new source of data: "Ancient DNA should be liberating to archeologists because with answers to these questions in reach, archeologists can get on with investigating what they have always been interested in, which is *why* these changes occurred."[48] As far as he is concerned, the information from ancient genomic data has done much more than inform our view of human history. He believes it has transformed, and will continue to revolutionize, our understanding of who we are, how we got here, and how we relate to one another today.

Although remarkable, the "genome revolution" was, and continues to be, exceedingly controversial.[49] As far as some archeologists were concerned, some geneticists were entirely too overenthusiastic about the explanatory power of genetic evidence, to the point they would exclude or downplay other forms of data from established disciplines like archeology, linguistics, and history. Sure enough, some archeologists felt some geneticists tended to embrace, intentionally or unintentionally, a reductionist mindset in terms of their choice of data (molecular data) as unsurpassed evidence for understanding human evolutionary history. Alexandra Ion, an archeologist at the University of Cambridge, pointed out the problem with the idea of ancient genetic or genomic data being a "holy grail" in the sense that this kind of data can always provide novel or better answers to old archeological questions. As an example, she drew on the case of King Richard III, who famously died in battle in the fifteenth century but whose remains and their whereabouts were left uncertain. In 2012, more than five hundred years after his death, researchers excavated a skeleton from underneath a car parking lot in Leicester, England, setting them off on a journey to identify a body that just might belong to the late king. In recounting the events, Ion outlined the ways in which researchers negotiated, and the media presented, the value of different lines of evidence from the genetic to the osteological, archeological, and historical. This multidisciplinary team of researchers knew their ability to extract and sequence genomic data from the ancient skeleton played a large role in identifying Richard III's body, but they also knew their confidence in this data depended on its correspondence to other evidence.

Although researchers may have recognized this, the media touted DNA as the real definitive proof, the evidence that solved the mystery. Ion expanded on this idea of negotiating evidence by drawing on more substantial works that used ancient molecular data to shed light on the Neolithic Revolution, a major period of transformation as people transitioned from a lifestyle as hunter-gatherers to farmers. She questioned whether the genetic data from the "hard sciences" was really being successfully integrated with the historical and cultural contexts of interest to archeologists (and with their traditional sources of evidence) in the so-called soft sciences. To be sure, archeologists were receptive to new methods and data, and many have forged strong, beneficial relationships with geneticists. Yet the idea that genetic data can always be appropriately integrated with archeological and historical evidence is a problem. As Ion argues, truly interdisciplinary research, when it comes to genetics and human history, is not easily achieved.[50]

The newfound access to ancient human DNA on a large scale and its increasing application to questions about human history has archeologists, as well as other scholars in the humanities, up in arms for a number of reasons. The issues and arguments are multifaceted, and the sides that geneticists and archeologists find themselves on are not wholly binary.[51] Given this, there are a number of worries about the hype around the genome revolution that has particularly affected ancient DNA research's disciplinary development going forward into the future.

One issue was not so much that archeologists denied the value of genetic data to illuminate answers to historical questions. Rather, the issue was overzealous confidence in genetic data to single-handedly answer big questions about human history through oversimplified and grandiose narratives about the past. In the broadest sense, some scholars—such as archeologists Rachel J. Crellin and Oliver J. T. Harris—identify this as the classic nature-culture binary, which they argue has informed much of ancient DNA research. They suggest that this not only is an inadequate understanding of the world but also leads geneticists, and even some arche-ologists, to favor genetic data, thus "placing archaeology and material culture in a secondary and subservient position."[52]

On one level, archeologists have been troubled by what seems to be geneticists' unbridled confidence in genetic evidence that competes with their own disciplinary methods and ways of knowing the past, be it material culture or ritual practices as documented through the archeological record. They argue that geneticists, archeologists, anthropologists, and historians alike should understand how their methods and data can complement rather than compete with one another. Their view is that DNA adds to the discussion while disciplines like archeology, history, and linguistics provide the context for the discussion in the first place.

On another level, archeologists, as well as historians, are even more concerned that geneticists' infringement on their territory may bring up unwelcome and outdated oversimplifications of human history. "Some archaeologists, however, worry that the molecular approach has robbed the field of nuance," writes Ewen Callaway in another *Nature* article. "They are concerned by sweeping DNA studies that they say make unwarranted, and even dangerous, assumptions about links between biology and culture."[53] Anthropologist Michael L. Blakey goes so far as to accuse genetics of biological determinism, namely the reduction of all cultural and societal phenomena to biological or genetic causes.[54] Such sentiments are far from a lone case of data envy but have much to do with larger cultural, societal, and political issues.

Although there are a number of research papers that illustrate such concerns for archeologists, Reich's book was a prime case in point and an easy target given both his prominent position in the field of ancient DNA research and the amount of praise he bestows on the field. In his book, Reich argues that ancient genomic research has the potential to study and discuss race on a scientific basis without necessarily being racist. He denies that his work is a form of scientific racism and instead argues that genetics actually transcends the social or cultural category of race concepts while dealing only with the biological facts of it.[55] Other scholars have pushed back on what they view to be a naive perception that scientists, even with the best of intentions, can simply separate the biological and cultural.[56] Some have argued more strongly, directly accusing Reich and colleagues of an outright racist ideology despite claims they avoid it.[57] Indeed, historians of science have pointed out the problems with such attempts to both divorce the biological from the cultural because there are always underlying assumptions, known or unknown, regarding issues of race, gender, ethnicity, and identity.[58] As the historian of science Jenny Reardon puts it, "As much as biologists have tried over the last several decades to constrict race to apolitical scientific purposes, the use of race is never neutral. It is always tied to questions with political and social salience."[59]

Further, historians of science, along with archeologists, have been arguably more alarmed by some geneticists' attempts to show how social-cultural categories easily correspond to genetic or other biological categories. They are alarmed when phenomena they perceive as cultural or social are reduced to something biological and when the biological explanation is given priority over all others. Archeologists have viewed much of the work in the field of ancient genomics, or at least the way geneticists talk about it, as intentionally or unintentionally resurrecting many concepts such as biological determinism that are not only outdated but morally problematic.[60] Recently, a host of archeologists, historians, and other scholars have highlighted the problems of trying to match biological with sociocultural concepts, especially as they relate to genetic ancestry companies, and the many epistemological and political risks that come with such practices.

To add to the controversy's complexity, archeologists have also been increasingly worried about the ways in which ancient DNA research is being conducted. Indeed, a large part of this angst over the use of DNA to answer big questions about human history has stemmed from the fact that labs like Reich's are growing into large-scale industrial operations managing the production and distribution of ancient DNA data. In his book, Reich is open about his objective to make "ancient DNA industrial" by transforming his lab into an "American-style genomics factory."[61] This science-turned-business philosophy has rubbed some researchers the wrong way. At one extreme, critics accuse Reich's lab of biocolonialism. Maria C. Ávila Arcos, a population geneticist at the International Laboratory for Human Genome Research in Mexico, notes that Reich's objective to industrialize the science carries insensitive undertones: "When one considers the social and historical context of the human populations that will be studied—many of which have been historically marginalized, colonized, and exploited—this statement becomes problematic." According to Ávila Arcos, "Such intentions could easily be perceived as a continuation of exploitation or biocolonialism." On this point, she argues that Reich's own use of an "unfortunate analogy further highlights the problem." She quotes Reich, who wrote in his book, "We are . . . like explorers in the late eighteenth century, sailing to every corner of the globe." As Ávila Arcos explains, "During the era to which Reich refers, European adventurers indeed collected samples from around the world, but these specimens were usually taken without the consent of, or regard for, the communities to whom they rightfully belonged."[62]

Others have argued that this colonialist attitude extends to more than the human populations being sampled and studied. In a feature published in the New York Times Magazine, writer Gideon Lewis-Kraus revealed that the big labs—Pääbo at Leipzig, Reich at Harvard, and Krause of Jena—are rumored to exercise power over some of the choicest human fossils.[63] These labs, Lewis-Kraus explains, enjoy access to money, technology, fossils, and top-tier scientific journals that makes it hard for smaller labs to compete or pressures them to collaborate. Consequently, many native fossils are said to be outsourced to the researchers of these bigger labs. According to

Lewis-Kraus, one scientist told him, "Certain geneticists see the rest of the world as the 19th-century colonialists saw Africa—as raw material opportunities and nothing else." This has contributed to an "atmosphere" of "anxiety and paranoia."[64] In fact, in his own conversations with archeologists and geneticists, Lewis-Kraus said that nearly all scientists asked for anonymity because they were concerned about professional backlash.

The search for DNA from ancient and extinct organisms has always been a high-profile research practice, and the recent rush to sequence ancient human genomes and rewrite human evolutionary history has only exacerbated the attention afforded to it by the media. According to interviewees, some feel the vast amount of data, the conclusions being made from this data, and the ever increasing celebrity status of the field is perhaps moving too fast for the field's own good. One interviewee likened the present state of the discipline to its early phase of research in the 1990s: "This research discipline has developed the way that all science—new scientific disciplines—develop, in that you have an initial, wonderful discovery, you have lots of hype and high expectations, and then you come down to it with a bump, and then you do the hard work of working out what it all means and what you can really do; what is realistic and what isn't. And that may take the next ten to twenty years of that research discipline." For this scientist, the community might currently be experiencing a second hype cycle: "I think with these next-generation sequencing techniques we have to do it all again; come down to it with a bump, and sort out what we can and can't do. So, I think it's cyclical" (Interviewee 5). What is distinctly different, however, about this phase of the discipline's development is that the implications for this sort of hype, and the consequences of such failed or misaligned expectations, are more serious than ever. In scientists' explicit confidence in their ability to rewrite human evolutionary history is the much more implicit claim that they can address the multifaceted political, cultural, and national identities woven into the history of people moving across the world and mixing with one another in the process. Indeed, the ethical stakes for this type of hype are profound.

SECOND HYPE CYCLE

The study of ancient DNA data had previously been limited to the study of mitochondrial DNA and sometimes nuclear DNA. Recently, however, the potential to sequence whole genomes via high-throughput sequencing technologies has allowed researchers to produce an increased amount of higher quality data (from several sequences to billions of sequences) that permits them to more accurately quantify contamination and therefore guarantee DNA authenticity. It has also allowed them to study the entire genomic makeup of an organism, similar to how modern genomes are analyzed, and this has provided more detailed answers to questions regarding phenotype, adaptation, and evolution, together with documenting when migration and gene flow events have occurred. As a result, researchers have recently reported that the "field" has "entered the new era of genomics and has provided valuable information when testing specific hypotheses related to the past."[65]

In interviews with ancient DNA researchers, some have suggested that the innovation of NGS of the early 2000s has ushered in a second hype cycle, much like the first hype cycle that the field experienced in the 1990s with the advent of PCR. Specifically, some feel the race for the first or oldest genomes is reminiscent of the race for the first or oldest DNA from ancient and extinct organisms. "I think a lot of the whole genome stuff," said one interviewee, "is just being driven by 'We're the first person to sequence the genome of extinct species X.' . . . And it's almost like the very early days of ancient DNA when you could get a *Nature* paper by saying, 'Ancient DNA recovered from extinct thylacine or quagga or Egyptian mummy or mammoth or whatever.'" As this scientist further explained, "It didn't really matter what the answer was. It was just the fact that you could do it. And I think that's possibly what's driving a lot of the ancient DNA community at the moment—is just again being the first to do something, not necessarily answering an intelligent question" (Interviewee 25).

Indeed, there seem to be parallels between the early years of ancient DNA research's disciplinary development in the heyday of the PCR era and the field's current optimism, with specific attention to the rhetoric of revolution surrounding the study of ancient humans across the world and over the centuries. One researcher, for example, said, "Several really big names in ancient DNA, they jumped onto the human train. I guess if they had all decided to work on megafauna, there would have been a bunch of papers already. It's coming, as soon as *Science* and *Nature* get tired of yet another ancient human genome paper. That's going to be the second wave in the next three to four years. You're going to start seeing *all* these extinct animals; their genomes sequenced, their population data sequenced." This researcher continued, "Scientists *know* that there are certain types of analyses that the media would go *more crazy* about than others, right? So, if you have the choice between . . . sequencing the genome of some random ancient human person and sequencing the genome of Richard III . . ., you'd go for sequencing Richard's genome because you know the media are going to go ape shit. You know ultimately that's going to lead to a higher likelihood of landing a grant" (Interviewee 38).

Furthermore, some feel the vast amount of data, the conclusions being made from this data, and the ever increasing celebrity status of the field is perhaps moving too fast for the field's own good. A leading scientist, for example, presented this perspective: "We have entered into another phase . . . where everybody thinks it's just so fucking amazing, right? . . . I think they will be super surprised in ten years from now—five or ten years from now—in terms of a lot of those claims need to be modified! And I think that we haven't by any means understood the limitations of what we are actually doing with genomics. And I think, you know, to be honest, I'm so surprised how the ancient genomics era has just been taken in by the anthropological community without questioning anything." The ancient genomics era was similar to the ancient genetics phase of the 1990s in terms of its exciting potential and exploratory nature. However, according to this interviewee, there were distinct differences too: "You can say the problem is not, I think, so much from the contamination. The problem is another kind now. . . . Now, it's the data analysis, really. It's the way you do the data analysis and it's the interpretations you are taking from that data analysis. . . . I can see already now that there's issues there, and I'm sure that there will be more to come. [Laughs] I think people will be pretty shocked" (Interviewee 7).

This is especially a problem because of the highly publicized nature of this line of research and the understandable tendency of media reporters to tell a clean and simple story that often does not do justice to the complexity of the research. As Ion argues in the case of King Richard III, for example, the media emphasized the role of DNA in identifying the skeleton over other lines of evidence that were equally, if not more, important. Anna Källén, an archeologist at Stockholm University, and colleagues found something similar in their analysis of the scientific research article and subsequent popularization of the famous Birka "warrior"—a skeleton discovered in a tenth-century burial chamber. When originally uncovered in the late 1870s, researchers assumed the skeleton belonged to a man, because the remains were buried with warrior equipment dating back to the Viking Age. More than a century after this initial discovery, scientists used ancient DNA data to determine that the skeleton was in fact female. In Källén and colleagues' study of media coverage on the recent findings, they found that the conclusions were communicated to the public by drawing on popular narratives and current political debates.[66] Likewise, archeologists Catherine Frieman (Australian National University) and Daniela Hofmann (University of Bergen) demonstrate how ancient DNA research on human population migrations across Europe have been exploited by far-right groups with racist, nationalistic, and political agendas. As Frieman and Hofmann argue, the blame for such misappropriation of ancient DNA research cannot be placed on one person or group, be it the media or the public. Nonetheless, scientists have a role to play in actively engaging with the implications (intended or unintended) of their research, especially given the intense press and public attention that accompanies it.[67]

Over the past three decades, the discipline has developed into what some see, despite this exploratory or experimental phase, as a more established practice in evolutionary biology. While there is certainly continuity regarding the interplay between science and the media from the PCR to the NGS eras, namely scientists' need for the press to maintain momentum to continue to be competitive in the field, there does seem to be a distinct difference. The hype around the search for ancient DNA is far less about resurrecting dinosaurs today, although this narrative certainly colors media reports and public discourses on new discoveries in the field. Rather, scientists' newfound technological and financial capacity to sequence entire genomes of archaic human specimens has led to expectations on another level, the ability of scientists to use ancient DNA data to rewrite our understanding of human evolutionary history. While the media plays a role in the hype, ancient DNA researchers are the ones explicitly promoting this promise. Ancient DNA researchers are indeed making more and farther-reaching claims about human origins, history, migrations, and admixture. Further, these claims also encompass centuries of historical, sociological, and cultural controversies over political and cultural identity. The hype around scientists' work as it relates to the insights ancient DNA data can provide for the study of human history has sweeping consequences for the broader public, sometimes with undesirable and potentially dangerous implications to empower outdated racist, nationalist, and political agendas.

CHAPTER EIGHT

Celebrity as Identity

DISCIPLINARY COHESION

After a thirty-year history in the media spotlight, and for the first time since the community schism over contamination in the early 2000s, three scientists—Erika Hagelberg, Michael Hofreiter, and Christine Keyser (a geneticist at the University of Strasbourg)—organized a landmark conference, "Ancient DNA: The First Three Decades" in November 2013. Much had changed since the 1980s, and this was a chance to reflect on past and present research. Hosted at the Royal Society in London, the conference consisted of nearly thirty talks over four days, the first half of which were open to the press and public and the second half that were privately presented at the Royal Society's Chicheley Hall. The meeting also resulted in a theme issue in the *Philosophical Transactions of the Royal Society of London,* featuring eighteen publications on the search for DNA from fossils.[1]

The meeting was more than a celebratory act. The commemoration also functioned as a way for scientists to reflect on and reinforce ancient DNA's place within evolutionary biology.[2] In the introduction to the special issue, Hagelberg, Hofreiter, and Keyser argued that the field was no longer a curiosity but now a credible practice: "In the past, a large number of ancient DNA studies were either purely technical, or one-off historical puzzles but, as we can see from the contributions to this *Theme Issue,* this is no longer the case, and ancient DNA researchers are now addressing a growing number of important scientific questions."[3] Leading up to this conference, a number of other papers suggested something similar. In an earlier article, for example, by Hofreiter and Michael Knapp, they claimed the search for DNA from fossils—once a "marginal discipline"—was on "its way into the centre of evolutionary biology."[4] The media echoed this sentiment too.[5] The ancient DNA community had come a long way, turning ideas of science fiction into a scientific reality. Overall, the conference highlighted this achievement as well as the struggles scientists had to overcome in the process.

Indeed, the search for DNA from fossils and its development into a discipline in its own right was not inevitable. Practitioners had to work at it as it required a merging of various disciplinary values, not without difficulty of course, to answer questions about the investigation of DNA from fossils and its applications to a range of biological and historical questions. Researchers from paleontology and archeology to molecular biology, microbiology, biochemistry, and genetics brought different skills and scientific and epistemic cultures to the table. Consequently, they were faced with the task of building the field from the ground up with consideration for, and conflict among, the various disciplinary expertise it attracted but also required.

The issue of contamination, despite the conflict it brought to the community, was paradoxically vital to its disciplinary cohesion, helping to define it as a distinct practice recognizable to other scientists as well as the public. In other words, contamination was a shared struggle across the community, a common problem requiring a common solution.[6] "There was definitely an ancient DNA community," said one interviewee. It was a community of "people with the same sorts of struggles who were applying it to a number of different questions" (Interviewee 30). Another researcher offered a similar perspective: "We were working in *completely* different areas and almost had no common ground of communication other than how difficult it was to get ancient DNA out of the sample and then whether PCR worked or not" (Interviewee 32). In fact, historian of science Elsbeth Bösl argues that the question of ancient DNA authenticity and the community's corresponding contamination concerns were an organizing principle that bound very different people together around a set of technical practices.[7] Contamination was, and to some extent still is, a common theme connecting scientists in the field.

While contamination was a main source of disciplinary cohesion, so was the celebrity that surrounded the field. Specifically, there were two ways celebrity influenced the field's growth over the decades. First, celebrity influenced the community's initial formation. *Jurassic Park* and the multimillion-dollar franchise that followed provided momentum behind the emerging practice, helping to marshal interest in terms of financial and organizational initiatives. Persistent publicity on behalf of media reporters—further nurtured by scientists, journal editors, and funding agencies—gave direction to ancient DNA research in its earliest, most vulnerable phase of development. "I think media has played a *huge* role in ancient DNA," said one scientist. "I think that it was intentionally used to play a big role in ancient DNA because, if you think about it, ancient DNA started as this field that was crazy!" As this scientist explained, "At the time, we didn't have the methods. We didn't have the know-how. . . . We needed that tie to build up to it and I think media was used to help generate interest and to maintain funding until we got to that point" (Interviewee 27).

This strategic and pragmatic use of media attention, especially in the early phases of research, is observable among other episodes in the history of science. Sociologist Elisabeth S. Clemmens, for example, detailed how scientists in the 1980s took advantage of the publicity surrounding controversy about what killed the dinosaurs as an opportunity for furthering discussion as well as research into the viability of competing hypotheses. The debate went public thanks to the asteroid impact hypothesis, and once it did, the media became a platform through which scientists communicated to one another across disciplines.[8] Clemmens argued that popular interest in this topic catalyzed communication among different disciplines, leading to increased research initiatives: "For astrophysicists, geologists, and geochemists, however, the link with a compelling question such as the death of the dinosaurs brought the promise of a new source of publicity, celebrity, and, perhaps, even greater funding." She also suggested that media attention helped scientists cross traditionally impermeable disciplinary boundaries: "Our usual image of the sciences is of a congeries of institutionally separate disciplines, each governed by a particular set of practices, professional norms, and cognitive orientations. But, as the impact debates graphically demonstrate, popular culture can serve as a matrix which fosters connections among disciplines that otherwise protect their institutional and intellectual autonomy."[9] As Clemmens made clear, a new scientific program can be initiated, then sustained, by popular interest.

Attention from outside the boundaries of what scientists see as standard science, such as the media, can in fact bridge disciplinary boundaries and sustain research development. For some interviewees, *Jurassic Park* was a critical component to the field's growth, and it may have helped sustain interest in ancient DNA research, particularly at a time when its credibility was challenged. One geneticist speculated that the field may not have evolved to the extent that it did without *Jurassic Park*'s influence: "If there was no *Jurassic Park*, I don't know how ancient DNA would be today," remarked this researcher. "I've said before I'm surprised that it actually survived the time with the PCR, and I think if there was no *Jurassic Park* it might not have come to anything at all" (Interviewee 34). Indeed, celebrity mattered at every stage of the discipline's development, even at a time when it was most contested, and at a time when celebrity was partly to blame (at least according to ancient DNA researchers). Even as researchers rejected the conclusions or implications of the *Jurassic Park* narrative, namely the long-term preservation of DNA and the ability to resurrect dinosaurs, they drew on the popularity of the novel and films to index the importance of the technical enterprise in which they were engaged.

Second, celebrity played a key role in ancient DNA research's overall identity as a scientific discipline. Consistent press and public recognition helped define the search for DNA from fossils as a research practice in its own right. This was especially important given the absence of an overarching theoretical framework, as well as in the absence of, or difficulty in obtaining, reliable financial or institutional support for research. Celebrity gave the field and the scientists within it a sense of legitimacy. Ancient DNA became a brand. Under this label, the corresponding conferences, newsletters, publications, headlines, and grants put a name to what was at first a far-fetched idea, thus helping to turn it into something exciting and worthy of investigation. Meanwhile, the resulting media attention and globally successful *Jurassic Park* franchise made it a recognizable and desirable research practice to the broader public. "It is good to have a brand that people can recognize and to show that what you do is different from other fields," explained a leading practitioner. "The term has certainly also been coined more by people who are trying to sell their science to the media as well" (Interviewee 12). To be clear, branding was far from superficial. It was purposeful and pragmatic. In a way, celebrity was a survival strategy for practitioners at a time when success was not necessarily a given. The preservation and extraction of DNA from ancient and extinct organisms had to be demonstrated. Obtaining press, public, and political support was important in doing so.

COMMUNITY CULTURE

Given the high risk of contamination, and high risk of failure because of the difficulty in demonstrating the authenticity and reproducibility of results, it seems surprising that so many individuals decided to enter the discipline. According to interviewees, however, the celebrity around the subject of ancient DNA research was a major reason for why some joined the search for DNA from fossils in the first place. Although there was a big risk of failure, there was also an equally big or even bigger reward should success be achieved. Not only did the media spotlight play a role in recruiting researchers into the practice but it also attracted the type of individual keen to take on its technical challenges and publicity potential. "I think . . . part of the reason it was fun was because you had to be slightly crazy in the 1990s to start getting into this field," said one biomolecular archeologist. "[My colleague] and I were slightly crazy, but we did it. Everybody was slightly crazy" (Interviewee 4). A second practitioner echoed this sentiment: "In the early days of the field, it was *ripe for failure*. Who's going to try and go out and get *DNA* from a *dinosaur*?" According to this interviewee, "If you wanted to set up a successful research career, and you wanted something safe, ancient DNA was *not your thing*. In the mid- to late 90s? No way!" (Interviewee 22). Indeed, there was a shared sense of risk in exploring such new territory that attracted a "quirky type of scientist" (Interviewee 4). It attracted a scientist with a "cowboy mentality"—those who were the "exact opposite of risk averse" (Interviewee 22).

The high-risk, high-reward nature of the search for DNA from fossils also produced a competitive community. "Everyone else was running towards the prize," said an early practitioner in the field. "One of the things that has annoyed me about ancient DNA . . . was that people were just running around to get the next sequence" (Interviewee 9). This race for the first and oldest DNA from some of the world's most charismatic creatures was not entirely unexpected. In fact, the race to be first in science is the result of what philosopher of science Michael Strevens calls the "priority rule." As he explains, the system in which scientific research operates recognizes and rewards those individuals—through publications, promotions, publicity, and funding—who come in first with a new theory, technology, or discovery.[10] In the field of ancient DNA research, the priority rule was all consuming. "You can sequence an extinct species for the first time only once," explained an interviewee. "So, the second headline is always smaller than the first one" (Interviewee 14). Consequently, competition was fierce in this field.

Within the community, researchers had their own hypotheses to explain the nature of the hunt for DNA from fossils. One evolutionary biologist said, "In ancient DNA it is relatively easy—getting a lot of attention—because it's a topic that naturally lends itself to media attention somehow. And I think that's one part that makes it so competitive and competition does not always bring the best out of people." On top of that, "many people are also very interested in media attention, so if it's important to be the first to get media attention, people do a lot of being 'the first.'" This scientist explained it further: "[A colleague] once mentioned it also has to do with the fact that you don't have to be particularly intelligent to be successful in ancient DNA— you have to be scrupulous to get the samples and to be faster than somebody else and to convince people to give you enough money." For example, "it's not like theoretical population genetics when you have to be a really good mathematician to make major contributions. So, it does not necessarily attract the *most* intelligent researchers, but it definitely attracts very competitive ones who are also *more* interested in presenting themselves to the media." According to this scientist, "that's also one thing in theoretical population genetics—you might get high citation rates, but no newspaper will write about it. Whereas if you can present the first mammoth genome, every newspaper will write about it" (Interviewee 15).

As interviewees noted, media attention was a cause and a consequence of competition among ancient DNA practitioners, so much so that it shaped their community culture. Competition for media attention often revolved around gaining access to fossils, especially the high-profile ones (dinosaurs, mammoths, or ancient humans). "A colleague of mine said it was almost like Gollum out of *Lord of the Rings*," explained a geneticist. "You have this *precious*. And maybe that's a throw over from anthropology but certainly power resided with those who could persuade people with bones" (Interviewee 21). The necessity but at the same time the rarity of fossils engendered much of this conflict. Another researcher added, "I know a couple of people who've said they've never come across anything quite as vicious and nasty as the ancient DNA field." When asked why competition was so fierce, this same scientist said, "Oh, because it's a small niche area. There's only so many really big questions you can answer or tackle. At least in the ancient DNA world, there are a lot of alpha-males kicking around who all want to basically not only own their territory but own the whole territory" (Interviewee 25). Disputes often occurred as researchers raced for access to specific samples that were especially newsworthy. "Actually, working in the ancient DNA field is often very difficult because there's so much competition over samples because the samples are very high-value, very sexy-type items," admitted one researcher (Interviewee 21).

Broadly, competition affected how different generations of researchers interacted with one another. One leading practitioner who entered the discipline at the height of the community competition in the late 1990s and early 2000s offered this opinion: "The generations before me—which I guess would be one to two generations of people before me—I always saw as being a very *nasty* community, extremely competitive, totally dominant. I think with my generation it has certainly improved." The aggressiveness caused some scientists to leave the field, but those who stayed tried to change the culture: "I remember that my generation . . . had always talked about [how] we were really tired of . . . the *aggressiveness* in the field, and from that perspective, I think we have had a different attitude. . . . We're still competing with each other and sometimes we're collaborating with each other." (Interviewee 7).

As the community expanded, the first generation of scientists from the early 1980s to early 1990s came into conflict with the second generation of researchers emerging at the turn of the century. "What's happened is that the children have killed their parents," one senior researcher explained. "There were these fairly average people getting into positions of power, but then their students were actually smarter than they were and so they kind of grew beyond their supervisors into an odd cycle of destruction" (Interviewee 9). For several students, their relationships with their supervisors were professionally or personally difficult. One practitioner who identified with the second generation said, "It's strange that the whole second generation is traumatized, somehow, by their Ph.D. supervisor, or by someone. I mean, we're not suffering anymore but we—all of us—had a period in our life where we suffered." For this practitioner, "That's what ties us together" (Interviewee 14). Another researcher of this generation made a similar statement: "If we think of that sort of generation then honestly I think the big change is that we are generally friends. . . . All those people—same generation—we get along pretty well together maybe because we all had pretty weird supervisors. [Laughs] So, yeah, that was a connection up front" (Interviewee 8).

In response to their strained relationships with first-generation supervisors, a number of second-generation scientists reacted by consciously creating a new atmosphere for themselves and their own future students: "There are two reactions from psychology," explained an interviewee. "When you're beaten, you beat back or you don't beat at all. So, I try not to beat at all. I try to treat my people better than my Ph.D. supervisor treated me. So, I very consciously thought about the social structure and the relation between people in my group because of that" (Interviewee 14). Another scientist added, "It just taught me the value of relationships. The samples are hard to get, but you can get them. The money is hard to get, but you can get it. People, though, as soon as you blow up relationships you have cut off access to money and samples and grants like you don't even know" (Interviewee 22). These practitioners' reactions to and reflections on their predecessors are examples of what the historian of science Joe Cain terms "patricide." "In the context of using history to construct heritage," Cain argues, "patricide is a systematic attempt to disconnect—to construct not relevance but *irrelevance*."[11] These second-generation scientists' efforts to break from past supervisor practices were systematic attempts to dissociate from the first generation, thus distinguishing themselves in order to create their own professional and personal identities within an already contested community.

IDENTITY CRISIS

In the introduction to the theme issue published in connection with the 2013 conference, the authors commented on what, for a lack of a better term, might be described as an identity crisis. Even in light of the practice's recent theoretical or technological developments, which scientists and media reporters felt indicated the maturing of a discipline, some researchers questioned whether ancient DNA research was even a discipline at all. "Despite these advances," wrote Hagelberg, Hofreiter, and Keyser, "ancient DNA research still has the feel of a young science. Some even doubt whether it is a field at all, or instead a collection of applications of molecular techniques to a variety of biological problems."[12] Interviews with researchers themselves reflected this dilemma too.

On one hand, some practitioners saw ancient DNA research as a field, a distinct discipline on its own. "I think it's a field," said one scientist. "You need lots of techniques in order to study it" but "it's not a technique" (Interviewee 23). Others argued for its status as a field too but one that was dependent on technology: "It's a field in itself, but it's strongly dependent on the techniques. The techniques are always shaping the field, and constraining it or expanding it. So, we will always be dependent on that" (Interviewee 49). Along this same line, a postdoctoral researcher proposed that the time, energy, and knowledge necessary to develop the methods and techniques were what made ancient DNA research a field on its own: "Ancient DNA—people develop methods to extract DNA and to preserve DNA or make it more efficient, cheaper. That's a field in itself and it has its own questions, but then you can use that branch to ask lots of questions" (Interviewee 51).

Conversely, a number of interviewees held that the search for DNA from fossils is more of a technique but also said that if it is a field, it is at least an unusual one: "I don't think ancient DNA is a research field as such. I'm going to be a bit—what's the word—provocative here. Or at least it's a very odd kind of research field. It's a technique." This interviewee shared a story to explain this point: "I can think of an ancient DNA researcher, but this person is by no means unique. In fact, it's quite common to people in that field. One minute they're working on the vegetational history of Siberia and the next minute they're working on what was the ethnicity of the first people to enter the New World, at the same time they're working on the phylogeny of camels." For this interviewee, this situation was distinctly different from other research agendas or approaches: "Now, you could not find that . . . in this institution. . . . There are people who devote their life to the phylogeny of algae. There are people who work on early mammal radiation. There are people who work on human evolution. This is because they've got some kind of drive because they're asking questions." But they quickly qualified this point: "I think the one thing that may be, might be said, in contradiction to what I've just said is that it *is* a specialist technique. You know, this . . . leading researcher who is working in all those areas is doing that because they have a *great* deal of expertise in extracting DNA out of different sources. That in itself is a skill" (Interviewee 3).

This disciplinary dilemma also extended to researchers' identities. The multidisciplinary and interdisciplinary nature of ancient DNA research, a practice that emerged from the interface of disciplines like paleontology, archeology, and genetics, problematized how scientists saw their professional identities. When asked about professional identity, one researcher replied, "This is a very good question and hard to answer. I give different answers depending on who asks me" (Interviewee 12). Other comments were comparable: "I'm a population geneticist or an evolutionary biologist or someone who tries to study human history with DNA or I don't know. It depends on the audience I'm talking to how I describe myself differently" (Interviewee 48). "I describe myself as a molecular evolutionist or a molecular archeologist. It depends on who I'm talking to and which project I'm working on at the time" (Interviewee 32). "I'm a jack-of-all-trades and master of none because I've done so many different things" (Interviewee 25).

Bösl also recognized the controversial nature of ancient DNA research's status as a new scientific field or technique used across other fields. As she notes, the status of DNA as source material for research questions, be they biological or historical, was heavily dependent on what question was being asked and who was doing the asking. For example, population geneticists work with both ancient and recent DNA and with both archeological specimens and living people today. Their data and the sources of the data span both space and time. In another instance, Bösl explains that in the study of bioarcheology, ancient DNA analyses are important but subordinate to other, more traditional sources of data and methods. Something similar can be said for the study of paleoanthropology and paleobotany in that ancient DNA analyses are valued as sources of data but are not a replacement for conventional morphological sources. In light of such diverse prioritization of ancient DNA data, Bösl acknowledges that the term "ancient DNA research" has been used by various researchers to loosely refer to a community of practitioners that emerged in the late 1980s, but that there are not hard and fast disciplinary boundaries that define or contain it.[13]

Ancient DNA's identity crisis over its status as a field, technique, or mix of the two has much to do with the fact that its multidisciplinary and interdisciplinary nature does not often or easily map onto traditional institutional frameworks for employment or research funding. Indeed, there is no one research journal, professorship, or clearly defined career pathway for this type of research. This means most researchers working with ancient DNA data are employed in a traditional archeology department or a genetics department, and in some cases they found themselves as researchers in both. One interviewee, for example, who was trained in anthropology and genetics reported feeling "firmly planted in both fields." At the same time, this interviewee—employed by two university departments (anthropology and genetics) to teach both social science and science undergraduates— admitted this overlap often left a sense of being spread too thin: "Sometimes it makes me feel like I'm not up in either" (Interviewee 30). This was more than a personal or professional problem. It was, and still is, a financial problem, especially for practitioners in the United States. Despite the fact that the search for DNA from fossils first attracted professional and popular attention in America, government grants for research have remained significantly lower than funding in Europe. A report in *Science* by Ann Gibbons notes, "While Europe forges ahead on a transformative technique, U.S. researchers struggle for funding." As Gibbons explains, "The interdisciplinary nature of the method is part of its power but also makes it prone to fall through the cracks in the U.S. system."[14] This tendency to fall through the cracks has had interesting implications for how ancient DNA researchers see the alleged discipline in which they work. In Europe, there are institutions that are better equipped to embrace the search for ancient DNA, working across traditional research boundaries in an effort to bridge knowledge and method gaps. Researchers such as Svante Pääbo and Johannes Krause, for example, with their training in biochemistry and evolutionary anthropology, feel at home at the Max Planck Institutes in Leipzig and Jena, respectively, because these institutions allow them the flexibility and funding to merge the two worlds of archeology and genetics.[15] But such a position is the exception, not the norm.

There were other reasons, suggested by scientists themselves, as to why their disciplinary status was under question. According to several interviewees, the innovation of NGS changed contamination concerns, making it less of an ever present issue and therefore making ancient DNA research less of a problem-centered science. "I think one of the biggest changes that's happened with ancient DNA is that it's really not even a field," said a younger, leading practitioner. "So, in the beginning, ancient DNA had its own thing, its own methods, it was all specialized. . . . It was all its *own separate world*." This changed as practitioners transitioned from the PCR and Sanger sequencing methodology to NGS: "And *now* it's really just genomics—applied to ancient samples. We don't have to adapt our data sets or adapt our methods or adapt our statistics anymore. We can just use the same statistics that everyone else is using in the entire field of genomics. . . . We can basically do anything that anyone else can do" (Interviewee 27). In other words, a formerly specialized technique was becoming more mainstream because contamination was less of a concern. "The voodoo is over," explained a geneticist (Interviewee 21). Without this problem, which defined the discipline for most of its history, scientists began to question the future of the field. One early practitioner, for example, suggested that a community was no longer necessary. "I think it's matured so much that there no longer is a community," this interviewee said, laughing. "You've got all these people using this technique—technologies—to answer questions and it's matured to the point where people from all these different areas don't really need to talk to one another" (Interviewee 24).

As much as contamination and the problems it posed for ancient DNA authenticity were a serious source of conflict for the community, they also created a strong source of unity. In an unpublished paper, two practitioners, Bernd Herrmann and Charles Greenblatt, offered this opinion of the future of the field: "It might be sad, but our community faces the fate of all pioneering communities. To a certain extent in our opinion the task is done. What now comes is the routinization within the subjects, wherever a[ncient] DNA technology is suitable and helpful in solving the problems in associated fields. The a[ncient] DNA research has been described as an interdisciplinary science. Here an uncertain future awaits us." As practitioners of an emerging practice with multidisciplinary interests and interactions, scientists were united through common problems concerning the preservation, extraction, and sequencing of DNA from fossils. In the late 1980s and early 1990s, contamination was a source of community cohesion as researchers discussed it in newsletters and debated it at conferences. But regardless of what side of the schism one was on, believers or non-believers, criteria of authenticity defined the discipline and colored researchers' memories of their histories and contributed to the shaping of their identities. Indeed, some argued that the hunt for ancient DNA was a method-based science defined around contamination concerns. Herrmann and Greenblatt remarked, "There is no common epistemological roof for a[ncient] DNA research but only a methodological one. It appears to us that most of the scholars in the field are *not* really interested in methodological and epistemological developments but in getting their specific questions answered. But in trying to do so, they had to turn themselves into method developers first and thus met a couple of people who were working on the same limitations, drawbacks, and pitfalls as they did." Further, "This is the true background for the a[ncient] DNA community, not a scientific program to solve a shared big question, and this may be one of our drawbacks."[16] Contamination was a core component of ancient DNA's social structure, but as it became less of an issue, a number of ancient DNA researchers questioned what the future of the field would be without it.

SCIENCE IN TRANSITION

The search for DNA from fossils emerged from the interface of disparate disciplines with distinctive research traditions and epistemic standards, from paleontology and archeology to epidemiology and molecular biology. Consequently, ancient DNA researchers had to create their own scientific culture from the intersection of many. The philosopher of science Peter Galison's notion of a "trading zone" is useful for understanding how practitioners from varied scientific backgrounds converged and went on to create a sort of unified community from what was a disunified beginning.[17] Drawing on a study of engineers and physicists in their collective work on particle detectors and radar, Galison used the notion of a trading zone to explain how the two successfully collaborated with one another from their dissimilar scientific paradigms. In the history of ancient DNA research, this shared interest in old molecules brought researchers together. Here, the idea of extracting DNA from fossils and using it to study evolutionary history was a boundary object that researchers rallied around and communicated across. But interest alone was not enough to establish a coherent community. As Bösl argues, those in search of DNA from fossils worked tirelessly, and sometimes unsuccessfully, to cross disciplinary boundaries and merge their divergent scientific, epistemic, and cultural differences.[18] This was especially true when scientists were faced with the extent to which certain sources of evidence, be it genetic, paleontological, archeological, or historical, could be combined and prioritized to make interpretations about the past, in particular about human history. In this trading zone, there was not always consensus but there was collaboration between, as well as competition among, researchers in the community that formed from it. In fact, there was a good deal of boundary-work going on as scientists tested and imposed limits through criteria of authenticity to try to standardize the research practice and their place as professionals within it.

Although criteria of authenticity, and the degree to which they are employed or enforced, were not a point of agreement, this was a shared concept and concern among ancient DNA researchers that affected the whole of the field. The search for DNA from fossils was an intensely technical and methodological challenge that required cooperation among scientists in order to overcome disciplinary boundaries and address the problem of contamination in regard to ancient DNA authenticity and reproducibility, both of which were seen as vital to the discipline's credibility. Over the years, contamination was a shared problem, and as such it became a defining part of this new group of scientists.

Yet celebrity played a part in the formation of this field too. Specifically, there were two ways in which press and public interest had a hand in ancient DNA's formation from the 1980s to today. First, popular interest helped to shape the practice of science by influencing the questions researchers asked, the funding they received, and the way they framed their research when communicating to wider popular and political audiences about its significance. Second, press and public recognition helped to define the search for DNA from fossils as a discipline in its own right. In the absence of an overarching theoretical framework and of consistent financial or institutional support, popular attention gave the field and the scientists within it a sort of status.

However, the introduction of NGS at the turn of the century changed the situation for the community. NGS offered an opportunity to overcome some of its most persistent technological challenges, namely the lack of data and problem of contamination. To be clear, NGS did not remove the possibility of contamination, but it did reframe the problem. Indeed, ancient samples had contaminating sequences, but practitioners were able to calculate the amount of contamination, permitting them to increase confidence in DNA authenticity. Without this pervasive problem of contamination, which defined the discipline for most of its history, interviewees questioned the future of the field.

In light of the field's general move from PCR to NGS, many interviewees argued that ancient DNA research would become more of a reliable and routine technique used as a tool in other fields, much like carbon-dating in the field of archeology. As a consequence, one interviewee predicted that the discipline would slowly but eventually "evaporate" (Interviewee 3). Another put the point more bluntly, going so far as to suggest the death of the discipline: "The death of ancient DNA has come about because no longer can you have a career as being someone who is good at getting ancient DNA out of old fossils. Now you have to actually understand that data." For this interviewee, the practice was becoming more mainstream as a means for answering biological, archeological, or historical questions: "The death is when you've got groups like David Reich's group or Svante [Pääbo's] group starting to just have ancient DNA as part of their group. Suddenly, it's just part of population genetics. There is no longer a discipline." As this interviewee further explained, "Ancient DNA was a discipline where you had to be specifically trained because levels of contamination were so high and the techniques were so poor that you needed specialist skills." As contamination became less of an issue, researchers turned to the expertise of statisticians and bioinformaticians: "You need to be someone whose head is in population genetics, and those people can now sweep up the ancient DNA data and interrogate it with modern data and do meaningful things with it. Consequently, the field of ancient DNA is dead." Crucially, this did not indicate its failure. Rather, some saw its death as its success: "It's not a death because it's done. It's because [it has] taken over. It moves into mainstream" (Interviewee 9). Another added, "The end of the ancient DNA society, as such, was because of its own success" (Interviewee 28).

New opportunities, as well as new obstacles, presented themselves. According to scientists in search of ancient genomes, population genetics was the new name of the game. One influential practitioner put it this way: "I've been waiting for the field to die out for twelve years now. I always considered ancient DNA to be mature when we wouldn't need an ancient DNA community anymore, but we would have the normal communities— ecology, evolution, archeology, whatever—and this would be applied as tools." This researcher quickly qualified this, explaining there was a need for a community, just a different type of community: "Since we're having a technical development, since we're having next-generation sequencing, since we're now working with analytical methods and software connected to next-generation sequencing, we're still having specific, or at least semi-specific, development in ancient DNA which is still to some extent legal-izing an ancient DNA community" (Interviewee 46). Expertise in mathematics, statistics, and bioinformatics was one way forward. As one geneticist said, "So, the future belongs to the geeks, not the Greeks, I suppose. [Laughs] It really does" (Interviewee 21). This new skill set would require researchers, particularly archeologists, to not necessarily become population geneticists in their own right but to understand what population geneticists are capable of, as well as their limitations, when reconstructing past population origins, migrations, and evolution.[19] The transition from PCR to NGS was not just a transition from one technology to another. For researchers, it represented a fundamental restructuring of the practice within wider scientific discourse.

CHAPTER NINE

Celebrity as Strategy

AN ANSWER IN SEARCH OF A QUESTION

Throughout the history of ancient DNA research, the practitioners working in this new field employed a number of strategies to aid their exploration of it. In their pursuit of DNA from fossils, scientists engaged in a *data-driven strategy* in terms of pursuing valuable samples, state-of-the-art technologies and techniques, and the molecular information (DNA) that could be recovered from accessible samples using those existing technologies. At the same time, they engaged in a *question-driven strategy* by asking and answering questions about the theoretical preservation and potential extraction of DNA from fossil material, and asking historical or biological questions about the organisms or populations under investigation. But researchers also adopted a further research strategy: a *celebrity-driven strategy*. With this strategy, scientists capitalized on publicity opportunities and created their own opportunities for attention too in order to communicate the excitement as well as value of this new line of research to professional, public, and political audiences. In doing so, scientists used the celebrity that surrounded the science (often by way of the charismatic creatures that were sampled and the potential to recover ancient genetic data from them) to drive the scientific and technological development of ancient DNA research in much the same way that the more traditional processes of data collecting or hypothesis testing did.

In marking the anniversary of the field's birth, the organizers of the "Ancient DNA: The First Three Decades" conference—Erika Hagelberg, Michael Hofreiter, and Christine Keyser—noted that the search for DNA from fossils had evolved beyond studies that were "purely technical" or "one-off historical puzzles." Today, ancient DNA researchers were "addressing a growing number of important scientific questions."[1] In other words, ancient DNA research was no longer solely technology-driven and sample-driven. At least according to these practitioners, they felt that it had matured as scientists were asking and answering more scientifically significant questions on a greater scope and scale. This change was the consequence of the introduction of NGS in the early 2000s and a subsequent shift from their technological dependence on PCR and Sanger sequencing. Indeed, for a number of interviewees, NGS freed researchers from PCR's constraints, thus allowing them to focus on the biological questions rather than the technological limitations: "For the first time in history," remarked one researcher, "I think we're not driven at all by the technology because the technology is permissive today. We are driven by the question we can answer with the technology. Well, it's not really that yet, but it's *close* to it" (Interviewee 8). According to another interviewee, it was more than the technology-driven nature of the practice that had passed: "I think that we are question-driven rather than sample-driven." For this interviewee, "now that all the low-hanging fruit have been picked it's more question-driven" (Interviewee 43). While scientists certainly recognized the role that technology and samples had played in the practice of ancient DNA research, they viewed this shift to a more question-driven approach as a mark of maturity. Maturity was desirable to scientists, especially amid credibility concerns.

In fact, reflecting on the field's thirty-year history, many other interviewees portrayed the practice in its early days as a primarily data-driven practice in terms of technology and samples. One interviewee even characterized the search for DNA from fossils as an answer looking for a question rather than a question looking for an answer (Interviewee 2). For some, the field was full of studies in which the answers, the DNA, seemed to supersede the questions that could be asked of the DNA. To be clear, this phrase, "an answer in search of a question," was a shorthand for studies that some scientists saw as being data-driven in approach. Such phrasing presupposed that ancient DNA sequences held relevant molecular information about the organisms they came from and their evolutionary history. Although true, actually accessing and then understanding the data required scientists to make sense of the DNA by analyzing, interpreting, and appropriately applying it to questions in evolutionary biology. In other words, the "answers" were not explicit in the sequences themselves. Scientists had to examine the data in order to make meaning of it. Nonetheless, scientists in the field often seemed to be data-driven in their approach in terms of the samples, the technology, and the molecular information that could be recovered from accessible samples using existing technologies.

This data-driven strategy, as other interviewees argued, was often coupled with a celebrity-driven approach to the search for ancient DNA. "They may have a research question," explained one paleobiologist, "but sometimes it's even *pre*-getting-a-research-question. It's like, 'Let's study these. Let's see if there's DNA in these fossils.'" This particular practitioner shared the following story about the priorities that went into their decision-making process for initiating research projects: "I remember one occasion when one of the well-known ancient DNA researchers said to me, 'What species should I study?' . . . It got to the point to where there would be Ph.D. students and you could see the supervisor thinking, 'What species hasn't anyone done yet? Nobody's done musk ox. Ok. You do musk ox.' *Without a very clear question.*" This data-driven approach seemed to be partly propelled by press and public interest in charismatic creatures that would yield data but also be likely to lead to high-impact publications: "I've seen several examples of 'let's blitz this species.' We give a Ph.D. student this species. They collect fossils from all over, they do the DNA, they draw up trees, and *then* they start to ask questions. . . . And then the supervisor is usually then looking for a high-impact angle," this interviewee said, laughing. "It is a slightly odd way of doing science" (Interviewee 3). This interviewee, as well as others, critically portrayed the science in its early years as a data-driven and even celebrity-driven practice.

To be clear, however, the search for DNA from fossils, even in its early era as a data-driven and celebrity-driven science, was in fact a question-driven one too. In the early studies, for example, researchers were indeed driven by questions—questions about the theoretical preservation and potential extraction of DNA from ancient and extinct organisms. And answering these questions was no small feat. Throughout the 1980s and 1990s, scientists confronted extreme technical challenges as they sought to discover what was possible regarding the preservation and extraction of DNA from ancient skins, tissues, and even bone. But for some studies, the questions also took on a biological bent. For example, the 1984 quagga study was initiated as a theoretical and technical challenge, but the specimen, *Equus quagga,* was specifically selected in order to test a hypothesis about the evolutionary history of an extinct species, one that was previously inconclusive based on fossil data alone.[2] Likewise, researchers working on one of the early studies to try to extract DNA from insects in ancient amber chose a particular termite specimen, *Mastotermes electrodominicus,* in order to test hypotheses of insect evolution and extinction.[3] In these cases, there was a biological question but it was secondary to the technical achievement of recovering DNA from the fossil in the first place.

In the early days, it was nearly necessary for the technical question to take precedence as the biological or historical question could not be answered without it. But precedence did not entail exclusion. In fact, in reality, data-driven, celebrity-driven, and question-driven approaches were not incompatible. Rather, they often went hand in hand. Such a mixed-method approach was not only possible but pragmatic. It was realistic, even inevitable. Sure enough, the use of one strategy was not exclusive of another. Nor was one strategy employed equally by all practitioners and to the same degree or frequency throughout the discipline's development. In other words, practitioners were prudent in choosing a combination of research strategies according to their circumstances, objectives, and the perceived pressures at the time.

While some scientists made arguments that they were entering into a more question-driven (and, in their view, a more mature) research era, other evidence—including other interviewee quotes—suggests otherwise. Indeed, a handful of interviewees argued that ancient DNA researchers were still very much driven by the available samples and technology. Indeed, according to several scientists, ancient DNA activity has continued to be an answer looking for a question. In 2015, for example, a team of researchers—including Morten E. Allentoft and Eske Willerslev in Copenhagen—sequenced 101 ancient human genomes ranging in date from A.D. 700 to 3000 B.C. with the ultimate goal of testing hypotheses about evolution and migration during a time when new tools and traditions had surfaced and spread across Eurasia.[4] Though question-driven, the project was also very much propelled by the samples and technology. Researchers went over the top to generate more genomes than necessary simply because they could.

Reporting for *Nature*, Ewen Callaway specifically spotlighted this research and took note of its data-driven approach. Callaway quoted Allentoft: "'We could have stopped at 80,' says Allentoft. But 'we thought, "Why the hell not? Let's go above 100."'" With NGS, the issue was no longer too little data but rather too much data. On this particular point, Callaway quoted Greger Larson at Oxford University: "'It's an interesting time, because the technology is moving faster than our ability to ask questions of it,' says Larson, whose lab has also amassed around 4,000 samples from ancient dogs and wolves to chart the origins of domestic dogs. 'Let's just sequence everything and ask questions later.'"[5] The trend continues today.[6]

Although there is certainly continuity between ancient DNA's data-driven past and present, the primary difference is that the situation has shifted from scientists having too little data to having too much data.[7] One practitioner, for example, commented on the impact this had on the research process: "We got some new genomes and it wasn't question driven anymore. We didn't have a look at those genomes because they were the key to a question, but [because] they were good samples and we could get whole genomes" (Interviewee 13). For a second scientist, this approach was a general consequence of new opportunities afforded by available technology as well as samples: "I think whenever a new technology comes on board there's a lot of 'Ta-da! Hey, we analyzed this stuff with this new technology.' And it's really driven by the labs that have access to the technology and the samples" (Interviewee 30).

Despite the fact that some of the earliest researchers were question-driven in their approach to the search for DNA from fossils, numerous interviewees in their memories of their history characterized this early research as being primarily data-driven. They tried to draw a line between what they viewed as a data-driven past and a more question-driven present in order to place some temporal and methodological boundaries around the evolution of their practice. In other words, this language of an answer looking for a question, rather than a question looking for an answer, was an extended episode of retrospective boundary-work. This language was a way in which scientists sought to compose a narrative of ancient DNA activity, intentionally or unintentionally, by drawing a line between its emergence and what some scientists see now as its more or less established status today. They engaged in this sort of retrospective boundary-work because they were concerned about their credibility within evolutionary biology, something that was challenged by both contamination concerns and what some viewed as disproportionate or undeserved media attention around sensational publications.

Crucially, the issue was not about whether ancient DNA research was science or non-science but whether it was a credible or non-credible approach to the study of evolution. The answer was far from simple as ancient DNA activity was tied up in a long history of scientific and popular expectations. Given that the search for DNA from ancient and extinct organisms had evolved into a discipline on a public platform, scientists felt they could not solely rely on technology or methodology in terms of protocols or verification as a way to draw lines between what they saw as reliable or less reliable work. They felt their public profile required a public response about the proper practice of ancient DNA research.[8] As a result, practitioners created criteria in the lab in response to contamination concerns, but they also built boundaries via rhetoric, especially through their memories of their history, in response to celebrity concerns.

Overall, demarcation mattered for scientists because the act of setting themselves and their work aside as reliable and rigorous signified relevance within evolutionary biology more broadly. Scientific maturity indicated authority, and this mattered for ancient DNA researchers coming out of a thirty-year history of credibility contests over contamination and celebrity concerns. To this end, the point is not to judge whether this data-driven and celebrity-driven approach was, or indeed still is, a positive or negative phenomenon in the world of ancient DNA research. Rather, the point is to highlight the fact that *scientists practiced science* in a way that was influenced by a want and need for both data and publicity, and that *they themselves* interpreted these influences as affecting the production of knowledge and their scientific status within evolutionary biology.

Scientists' efforts to distinguish credible from less credible research was not unusual. In the history and philosophy of science, demarcation is a well-known topic of debate.[9] In fact, the demarcation issue has a long history of heated discussions about the more or less correct ways, and even wrong ways, of practicing science and what gets to count as proper science.[10] The famous philosopher Karl Popper argued that science could be distinguished from non-science by the fact that a given hypothesis could be tested and proved false. Popper's criterion of falsifiability, also known as the testability or refutability of a hypothesis or theory, still holds strong as a benchmark for demarcating science from non-science, as well as good science from bad science.[11] In fact, hypothesis-testing as a criterion for proper scientific practice had often been given a privileged position over other methods of inquiry.[12] Indeed, with the rise of big data across the scientific disciplines, and what many refer to as a new mode of data-driven scientific research, the place of and preference for hypothesis-driven inquiry has come back into the spotlight.[13]

QUESTION IN SEARCH OF AN ANSWER

Philosophers today are increasingly interested in this phenomenon of data-driven science. In 2012, for example, a group of scholars approached this topic with an intent to identify its characteristics as well as its causes and consequences for the production of scientific knowledge. They were also interested in trying to understand the role that hypotheses and theories played in this sort of methodology. According to their studies, they found that data-driven sciences often value the process of induction from given data as a legitimate approach to scientific inference and the role of technology as a means of analyzing and then extracting significant patterns from the data. In all of this, philosophers have asked a further question of this particular phenomenon: does this data-driven approach constitute a novel approach to scientific inquiry or does it share similarities with past research practices?

In his commentary on a selection of these papers, Bruno J. Strasser looked for overall similarities and differences among data-driven sciences over the past few centuries.[14] As far as he is concerned, early natural history "wonder cabinets"—collections of oddities from geological, historical, or religious relics—were not so dissimilar from "electronic databases" of today. According to Strasser, "Renaissance naturalists were no less inundated with new information than our contemporaries." Indeed, "The expansion of travel, epitomized by the discovery of the New World, exposed European naturalists to new facts that did not fit into the systems of knowledge inherited from the Greeks and Romans."[15] Through a study of Carl Linnaeus (an eighteenth-century physician and botanist from Sweden), scholars Staffan Müller-Wille and Isabelle Charmantier specifically spotlighted the data-driven nature of natural history in terms of the various strategies Linnaeus employed to organize and analyze what can be called an "information overload" of new species data. First, these authors draw attention to the fact that

Linnaeus used new tools—such as dichotomous diagrams, files, and indexes—to help control the amount of data under study. Interestingly, however, these tools, initially intended to control the amount of data, facilitated an influx of data. The authors point out that in the midst of this data deluge, Linnaeus attempted to make sense of the information by generating a hypothesis, namely the genus concept as a distinct category, as a further means for organizing the information, then classifying and comparing organisms accordingly.[16] Strasser puts the point this way: "In other words, Linnaeus may have been driven by his data, but his approach was not exclusively data-driven."[17] This example showcases ways in which data-driven inquiry may not be so new to contemporary scientific and technological practices. Rather, natural history has a long tradition of producing, then dealing with, information overload. Further, natural historians, like contemporary scientists, were also open to pursuing mixed research methods such as data-gathering and hypothesis-testing to make sense of the world around them.

Past and present data-driven practices have distinct differences. For Strasser, three features set the contemporary data-driven sciences apart from former natural history practices: (1) the data analysis today is done by researchers from disciplinary backgrounds different from the individuals who produced it, (2) the data analysis depends on the use and understanding of statistical tools, and (3) the data is primarily generated from inside the lab and not the field, as was typical of previous natural history practices. Strasser also attempts to explain why so many scientists view data-driven inquiry today as uniquely overwhelming and even revolutionary: "To conclude, it is mainly because the experimental sciences took the upper hand over natural history in the late nineteenth century and have since come to dominate the public perception of science that data-driven research is now perceived as a novel feature of twenty-first century science." Yet, historically minded case studies demonstrate that this data deluge is nothing new: "Natural history had been 'data-driven' for many centuries before the proponents of postgenomics approaches and systems biology began to claim the radical novelty of their methods."[18]

In her book *Data-Centric Biology: A Philosophical Study*, Sabina Leonelli makes an additional observation about the data-driven sciences of today. According to Leonelli, data-driven sciences are not necessarily interesting because they are data-driven but because of the social, organizational, and institutional structuring required to produce, then analyze, the massive amounts of data that are a part of the practice.[19] In her case study, she focuses on plant system databases in model organism biology to map out what she calls data-journeys. Here, she is interested in outlining the ways in which researchers across the board work together to collect, integrate, analyze, and share various sources of data that eventually will be used for different scientific purposes. Leonelli's focus on data-driven science is much more about the process than the product.

Likewise, ancient DNA researchers seem to be making moves to operate under similar large-scale organizational systems. Ancient DNA researchers, in light of new whole-genome sequencing technologies and techniques, face new challenges and are trying to generate a new kind of institutional infrastructure in response. Indeed, ancient DNA researchers are finding that a whole host of resources are required to go from a sample to a sequence to meaningful scientific analysis in a reasonable time frame. In response to opportunities offered by technology, some scientists are responding by building large-scale business-like operations that oversee the production and distribution of ancient DNA data. Svante Pääbo's lab at the Max Plank Institute for Evolutionary Anthropology, Eske Willerslev's lab at the Center for GeoGenetics in the Natural History Museum of Denmark, and David Reich's lab in the Department of Genetics at Harvard University are examples of the industrial operation that ancient DNA research can require. Much of this go-big-or-go-home approach to the search for DNA from fossils is of scientists' own doing.

Most recently and obviously, new whole-genome sequencing technologies have pushed ancient DNA researchers to seek additional skills in statistics, bioinformatics, and population genetics in order to analyze the massive amounts of data that can now be extracted from hundreds of samples. Crucially, ancient DNA researchers are much more than a user community of the machinery. They are committed to developing new methods that can be used in the lab to optimize the extraction and sequencing processes. In other words, although new technologies and techniques are critical to ancient DNA activity and the extent to which data can be made available and analyzed, practitioners do not just draw on developments in other fields but instead are active in adapting these innovations for their own purposes. Ancient DNA requires manipulation and management of data because the nature of ancient DNA is not the same as that of modern DNA. The extraction, sequencing, and analysis of degraded and damaged DNA requires a specialist skill set to understand the biochemistry of DNA damage and to correctly infer how differences between sequences relate to differences among individuals and populations over time.[20]

This shift to ancient DNA research as an industrial operation highlights the changing ways the field has become or might be even more data-driven in the future. Over a thirty-year period, the practice evolved from an initial effort to try to extract DNA from fossil material with the technology of PCR to a big-risk, big-reward initiative as practitioners set out to maximize the amount of genomic data that could be produced using NGS. In fact, some labs sought to turn the science of ancient DNA research into a truly large-scale, industrial, and automated process.[21]

Data-driven approaches have been and continue to be a principal part of the search for DNA from fossils. Traditionally, philosophers of science and scientists themselves have tried to divide scientific inquiry into data-driven versus hypothesis-driven. This binary view, however, is changing as scholars bring attention to the fact that data-driven research is often pursued in combination with other modes of scientific inquiry. For example, a number of scholars have addressed the role of exploratory experimentation in the data-driven sciences.[22] In his case study of systems biology, Ulrich Krohs makes the point that all science does not need to be hypothesis-driven at all times, nor is it even practical. Krohs, arguing against the classical conception that the goal of experimentation is to test hypotheses, suggests that "other modes of experimentation," such as the "searching mode of exploratory experimentation" as well as "data driven research," can be considered as "serious epistemic strategies, besides, and in combination with, hypothesis driven research."[23] Other philosophers also argue for the need to make more room for an interplay of approaches when it comes to understanding the process and practice of science. As philosopher of biology Maureen A. O'Malley writes, "It is possible that theory-driven hypothesis testing has been conceived of by scientists, science funders and philosophers in a way that does not exist in practice (and never has), and that it is closer to and involves more interplay with exploratory experimentation as well as natural history experimentation than we have tended to think."[24] As the history of ancient DNA research has demonstrated, a celebrity-driven strategy—often in combination with data- and question-driven approaches too—has been a frequent and fruitful driver of the field.

The Neanderthal Genome Project is just one exemplary case of this. In 2006, shortly after the introduction of NGS and its initial application to a number of ancient DNA studies, Pääbo and the MPIEVA in Leipzig, along with 454 Life Sciences Corporation, announced they would be the first to attempt to sequence the entire Neanderthal genome. Through an orchestrated press conference and press release, they announced they would accomplish the task in a mere two years.[25] From the outset, the Neanderthal Genome Project was a substantial, and quite intentional, media production that played on the technology, celebrity, and research impact of the findings eventually to be produced. This was not unprecedented. For example, science studies scholar Stephen Hilgartner argues for the increasingly intense "media-orientation" of genome researchers during the days of the Human Genome Project (HGP). For Hilgartner, "science-media coupling" was "strategic interaction." He suggests that these genome researchers turned to the media in the face of competition in the race to sequence the human genome. Indeed, they were very conscious of their behavior and orientation toward the media. As Hilgartner explains, "HGP leaders, for their part, arguably did what the managers of any enterprise would" regarding their decisions to "react strategically to emerging events" and "tailor media messages that would defend their legitimacy."[26]

SCIENCE IN PRACTICE

In their recollections of their history of ancient DNA research, scientists attempted to draw a distinction between what they saw as a data-driven and celebrity-driven phase of research versus a more question-driven methodology. The sometimes derogatory or dismissive comments by some interviewees about earlier practitioners, or even practitioners today, as scientists merely chasing samples, technology, and even celebrity, can be viewed as an extended episode of boundary-work, an attempt to create rhetorical and epistemological distinctions between the field's past and present. Their views of how others in the field utilized certain strategies, either exclusively, appropriately, or inappropriately, is a matter of opinion for what they viewed to be the proper processes and practices of science, whatever those may be. In doing so, the individuals interviewed were aligning themselves with one scientific approach over another. According to some of them, being question-driven rather than sample-, technology-, or celebrity-driven was a hallmark of scientific maturity.

Although practitioners' boundary-building was sociologically important for establishing their identity and authority within the scientific community, this boundary-work was also naive given that philosophers today no longer necessarily see a distinction between data-driven and hypothesis-driven research, and scientists do not actually practice science in such a binary way. Instead, more recent philosophical viewpoints argue for the need to make room for more forms of inquiry in scientific practice. In line with this, a celebrity-driven strategy—as clearly and consistently utilized by ancient DNA researchers themselves—can be considered a "serious epistemic strategy" that practitioners, as well as editors and funders, employ when making choices about research agendas, publication acceptance, and grant funding. Certainly, not all research was guided by (or need be guided by) its potential to attract popular interest, but in this history, the celebrity that surrounded the science of ancient DNA research was a crucial consideration behind researchers' decisions which influenced their process of data-gathering and hypothesis-testing.

This either-or characterization by interviewees of a data- versus question- versus celebrity-driven approach to the search for DNA from fossils is to some extent misleading. In fact, ancient DNA researchers used a variety of approaches, sometimes foregrounding the appeal of celebrity over the question under study, other times prioritizing the technology over the celebrity. In reality, scientists—ancient DNA researchers included—use multiple approaches simultaneously and iteratively, assessing the achievability of research results against accessibility to technology, samples, and funding against the prestige and publicity they could gain. Indeed, at times in the discipline's development, especially with the introduction of a new technology, scientists prioritized the use of this technology to the extent that sometimes it did more explanatory work than the celebrity of it. At other times, scientists emphasized or downplayed the celebrity of their work, all depending on what they wanted to accomplish. Regardless, ancient DNA researchers were opportunistic and pragmatic, even if some found it somewhat disagreeable.

CHAPTER TEN

Jurassic Park Effect

DE-EXTINCTION

Since the birth of the field in the early 1980s, the search for DNA from fossils has been inextricably linked to the idea of resurrecting extinct species, dinosaurs in particular. This connection, forged by scientists (as well as media reporters), was more firmly established and then reinforced by the continued public interest in the developing discipline thanks to the global success of the *Jurassic Park* franchise in the 1990s. During this decade, a number of scientists perpetuated the rhetoric of resurrection to generate media attention, even if the idea of bringing extinct creatures back to life was a far-fetched one.

Today, *Jurassic Park*'s legacy is ongoing, and scientists who have any association with the search for DNA from fossils are repeatedly asked the answer to this one question: Can we bring back a dinosaur? Interestingly, the majority of ancient DNA researchers have very little to do with the science of species resurrection—also known as de-extinction—mainly because the recovery and application of ancient DNA data is only one plausible point in a series of steps necessary to achieve such a grand task. However, many ancient DNA researchers certainly have opinions about whether bringing extinct creatures back to life could, or should, be done.

In the ancient DNA community, attitudes toward de-extinction are highly varied. Many interviewees argued that the task still requires incredible technological and biological improvements, as well as philosophical, environmental, and ethical considerations. Some even argued that de-extinction is all sensation and no science. Consequently, they actively avoid any association with it (Interviewees 2, 48). Others outright opposed the idea. One described de-extinction as a "freak idea" (Interviewee 37). Another called it "ludicrous" (Interviewee 2). Others referred to researchers involved in de-extinction efforts as "weird," "crazy," or "mad" (Interviewees 30, 18, 5). Additionally, some researchers were wholly oblivious to the de-extinction movement simply because it was in no way related to their own work (Interviewees 1, 14, 21). However, there were some scientists who were not so quick to dismiss the idea of bringing extinct creatures back to life, mainly because they were much more informed about the most recent advancements that could make such an arcane idea a reality (Interviewees 6, 15). Indeed, new technologies and techniques—as well as significant organizational initiatives—over the last five to ten years have made a number of ancient DNA researchers reconsider resurrection as a legitimate prospect.

In 2013, for example, the National Geographic Society in Washington, D.C., held the first public conference on the science and ethics of "De-Extinction." During this one-day event, more than twenty people of varying expertise across the world addressed the scientific, technological, political, and ethical implications of bringing back extinct species. Co-hosted by National Geographic Society, TED, and Revive & Restore (a nonprofit founded by Stewart Brand and Ryan Phelan for studying biodiversity and reviving endangered or extinct species), the meeting marked the first time the topic of de-extinction was openly and seriously discussed.[1]

The conference, called TEDxDeExtinction, included scientists in fields ranging from paleontology and genetics to conservation biology, ecology, and synthetic biology. Some scientists from the ancient DNA community were present too. Hendrik Poinar from McMaster University in Ontario, and Beth Shapiro at the University of California, Santa Cruz, both spoke on the possibilities but extreme difficulties of mammoth de-extinction.[2] Both Poinar and Shapiro—former students of Alan Cooper at Oxford—had since earned respectable reputations as leaders in the field of ancient DNA research. Although they entertained the idea of de-extinction, they were much more reserved, and to some extent critical, about its implications. Ben Novak, a former student of Poinar's and now a collaborator with Shapiro, also presented his project, sponsored by Revive & Restore, to resurrect the passenger pigeon.[3] In stark contrast to

Cover of *National Geographic* featuring the TEDxDeExtinction conference with an article by Carl Zimmer, "Bringing Them Back to Life" (April 2013). The article's subtitle reads, "The revival of extinct species is no longer a fantasy. But is it a good idea?" (Reprinted by permission from *National Geographic*)

Poinar's and Shapiro's talks, Novak eagerly and confidently argued that de-extinction will indeed be possible and have a positive impact on the environment. Indeed, achieving such a huge task has become his personal research mission. Regardless of their outlooks on the future of de-extinction, these practitioners realized that the idea of species resurrection could no longer be dismissed out of hand. As such, it had to be addressed upfront.

Recently, de-extinction has evolved from an idea into an actual research endeavor thanks to technology like NGS and CRISPR/Cas9, a technique allowing researchers to edit genomes by cutting out specific DNA sequences and inserting other sequences of interest. "In the good old days, you could conveniently say, 'You can't clone a mammoth.' And that was the end of it," remarked an interviewee. "But with NGS, it's got a bit difficult now" (Interviewee 2). The newfound ability to sequence whole genomes coupled with CRISPR—the 2015 "breakthrough" of the year according to *Science*—was the technical impetus that caused practitioners to reconsider species resurrection.[4] Highly accomplished, respected scientists such as George Church, a geneticist at Harvard Medical School, take de-extinction seriously, talking about it publicly and even heading efforts to investigate its viability. One researcher recalled, "George Church is what took it away from being crazy . . . to credible. Whenever he says anything everyone listens because he really knows what he's talking about." In many ways, de-extinction demanded a genuine reevaluation. "It's dangerous to say that the de-extinction people are crazy people. It *used* to be crazy people," explained this same scientist. "The reason I got interested is because I suddenly saw serious people in it. So, there are certainly crazy de-extinction people, but there appear to be very credible de-extinction people" (Interviewee 6).

Today there are multiple approaches that scientists suggest for bringing extinct creatures back to life, from back-breeding and cloning to genetic engineering.[5] The South African taxidermist Reinhold Rau's project to resurrect the quagga is one example of back-breeding, an approach that implements selective breeding processes with living zebras to recreate the quagga's unique striping pattern. Rau died in 2006, but before his death he witnessed the birth of Henry, the result of three generations of back-breeding and the first phenotypically representative relative of the quagga in over a century.[6] In addition to back-breeding, cloning offers another option for resurrection research. In 2000, the Pyrenean ibex, commonly called the bucardo, went extinct. In 2009, a team tried to bring it back via cloning, and out of hundreds of embryos, one baby bucardo was born but died ten minutes later because of developmental lung deficiencies.[7] Additionally, Revive & Restore has ongoing opportunities using genetic engineering to make extant organisms more like extinct ones, such as Novak's and Shapiro's efforts to bring back the passenger pigeon.[8] A second study headed by Church at Harvard is also experimenting with genetic engineering through genome editing to resurrect the woolly mammoth.[9] Other scientists, like paleontologist Jack Horner, have set out to make the world's first "chicken-o-saurus," via experiments with reverse genetic engineering research to make a dinosaur-look-a-like out of a chicken.[10]

However, scientists' use of technologies also depends on biological and philosophical considerations. For example, the question of what makes a species a "species" is a long debated one in the history and philosophy of biology.[11] De-extinction raises this question again but from a very different view. In 2013, Stanford Law School hosted a meeting titled "De-Extinction: Ethics, Law, and Politics." At this gathering, experts from lawyer Hank Greely to Steward Brand of Revive & Restore "defined true de-extinction" as "when a complete organism from an extinct species is brought back to life." The definition was far from perfect as it begged the question of species continuity or authenticity.[12] For example, interviewees say epigenetics, the external and environmental factors that play a part in organism development, are incredibly important to consider. One researcher remarked, "If you were to recreate a mammoth, and you were to release a mammoth—from an elephant—then it may not have any of the mammoth epigenetic changes because it's coming out of an elephant, and that could be massively important because we know epigenetics are so important now" (Interviewee 6). Others say these biological complications have philosophical implications regarding the difference between a mammoth and what one interviewee referred to as a "pseudo-mammoth."[13] A second scientist commented, "You can mammoth-ize it" but "it will never become *the* mammoth" (Interviewee 37). A leading ancient DNA researcher echoed this point: "It's not a de-extinction" but "a sort of 'frankensteinization' of life" (Interviewee 8).

In contrast, a postdoctoral researcher maintained that defining de-extinction requires a much more nuanced perspective: "It's a *very* interesting problem to start with, and again it comes down to semantics. How do you define a passenger pigeon? Is it once you've changed enough genes? Do you need to change everything so that you have a completely identical genome to the bird that you got the genome from?" According to this interviewee, "Even if it's genetically identical to a passenger pigeon it's probably not going to be environmentally and ecologically [the same]" (Interviewee 52). Another scientist argued that in the case of the passenger pigeon project, authenticity would be determined by functionality: "We don't necessarily want to duplicate a passenger pigeon, but we want to duplicate its ecology. We're just focused on the traits that make passenger pigeon ecology, not necessarily what makes the organism 100% 'it'" (Interviewee 45). Revive &

Restore, home to the passenger pigeon project, says the same for their project to bring back the mammoth.

Crucially, there are also arguments for de-extinction's scientific significance, as several scientists in the ancient DNA community are taking the time to educate themselves about its potential. In *How to Clone a Mammoth,* Beth Shapiro presents the first comprehensive account of the science and science fiction of de-extinction.[14] She explores the feasibility of de-extinction by outlining the series of steps and obstacles to overcome to make it a reality. She also considers the possible payoff of de-extinction, arguing that the ultimate goal is not necessarily the resurrection of a long-lost species. Rather, the goal is the revitalization and stabilization of ecosystems. For example, reviving the mammoth, or an elephant with mammoth-like traits, and restoring it to colder climates could reintroduce the growth of the grasslands that were once an essential element of the tundra ecosystem before the mammoth's extinction. A journalist reporting for *Newsweek* explained it in these terms: "For all its charisma, the mammoth is just part of a grand new strategy to restore long-gone megafauna. Scientists call it rewilding. The idea goes hand in hand with new thinking about the relationship between humans and nature—namely, that even the earliest civilizations had what we might think of as an unnatural impact on the natural world around them."[15] In these terms, de-extinction can be understood as part of a holistic effort under the name of conservation biology.

One of the grandest rewilding attempts is being pursued by two Russian scientists, Sergey Zimov and Nikita Zimov. At the turn of the century, they announced the founding of Pleistocene Park, a nature reserve in northeastern Siberia dedicated to the regrowth of the subarctic grassland ecosystem that could one day be home to the woolly mammoth, which thrived tens of thousands of years ago.[16] *Newsweek* called it "A Real-Life Jurassic Park."[17] Despite this rebranding of resurrection as conservation biology with a twist, the idea of reviving extinct species and recreating the environment in which they once lived is still a mix of science combined with a big dose of spectacle.

SCIENCE AND SPECTACLE

For a number of ancient DNA researchers, de-extinction was a moral issue and one that needed to be rectified. Paleontologist Michael Archer, the director of the Australian Museum in Sydney, for example, was a primary champion of this view, arguing that we have a moral obligation to bring back extinct species such as the thylacine because we—through human population and predation increase—were the cause of their demise more than a century ago.[18] Indeed, Archer was the mastermind behind the "Thylacine Cloning Project"—a bold and pioneering effort to resurrect the thylacine by extracting and sequencing DNA from a hundred-year-old pup pickled and preserved in a jar on the shelf of the museum. In 2000, at the height of the hype but also of skepticism for the search for DNA from fossils, the Australian Museum broadcasted the Thylacine Cloning Project and promised to meet its goal within the next twenty years.[19] Naturally, the project's debut made national and international news.[20]

Although Archer and the Thylacine Cloning Project were the subject of much enthusiasm and media interest, they were also the target of criticism from colleagues. Unsurprisingly, fellow scientists were less than amused, protesting the project on grounds that it was more spectacle than science. One argued it was nothing more than a strategy to attract press and public attention. "I've just said bollocks all the way through," fumed one senior researcher. "I get very upset. I could talk myself into a bit of a frenzy over it because I think it's so misguided. It isn't going to happen, in my opinion. It's cheap and sleazy in terms of the way that it attracts media attention and the public." As far as critical colleagues were concerned, the technological and biological hurdles were much too high, but they also took issue with the way the project was presented because, as they argued, the project was all hype and premature publicity with little to no technological or scientific evidence to support its feasibility. "The thylacine is a perfect example. 'We're doing it! It's almost there! It's going to be here in a year or two!' It's a fucking snake oil salesman" (Interviewee 32).

As science studies scholar Amy Lynn Fletcher perceptively notes, the debate around the thylacine's resurrection was chiefly a controversy over control for how the project should be framed and interpreted, as either science or spectacle, across mass media. "When it launched the thylacine project," argues Fletcher, "the Australian Museum walked out on the unstable precipice of 'paleogenomics as science' versus 'paleogenomics as spectacle.'"[21] Researchers used the media spotlight to promote the project, but in doing so, as they realized, it became a source of destabilization. According to Fletcher, the museum and its proponents found it difficult to control the story. While they touted it as a scientific and technological innovation, opponents accused it of being nothing more than a spectacle.

In addition to Archer's thylacine cloning project, there have been a host of other resurrection efforts that captured public interest and generated professional criticism. Today, the rhetoric of resurrection is still closely connected to the search for ancient DNA, but there has been a subtle shift in focus. Questions are less about dinosaur resurrection and more about mammoth de-extinction. In 2006, for example, scientists sequenced 13 million base-pairs of DNA from a 28,000-year-old woolly mammoth, demonstrating the powerful potential of NGS as applied to fossils. According to the press, this was a step closer to generating whole genomes, which also meant a step closer to bringing long-lost creatures, such as the mammoth, back to life. *Newsweek* reported this research, explaining, "The scientists, in other words, had managed to assemble half the woolly-mammoth genome; they claimed that in three years they could finish the job. That would put scientists within striking distance of an even greater feat: repopulating the earth with creatures that vanished ages ago."[22]

Two years later, in 2008, another lab sequenced the nuclear genome of the woolly mammoth.[23] The *New York Times* announced, "Scientists are talking for the first time about the old idea of resurrecting extinct species as if this staple of science fiction is a realistic possibility, saying that a living mammoth could perhaps be regenerated for as little as $10 million." Reporter Nicholas Wade explained that the feat was not so simple, but one practitioner, Stephen Schuster at Pennsylvania State University, made it seem within reach. According to Wade's reporting, "There is no present way to synthesize a genome-size chunk of mammoth DNA, let alone to develop it into a whole animal. But Dr. Schuster said a shortcut would be to modify the genome of an elephant's cell at the 400,000 or more sites necessary to make it resemble a mammoth's genome. The cell could be converted into an embryo and brought to term by an elephant, a project he estimated would cost some $10 million." Wade further quoted Schuster, "'This is something that could work, though it will be tedious and expensive.'"[24] One researcher recalled this comment and noted how it appealed to the press: "The first mammoth genome . . . Stephen Schuster . . . provided one of those quotes to the media that they love. He said something like, 'Give me a few million euros or dollars and I will recreate the mammoth for you'" (Interviewee 37). Even if this were possible, sooner or later, it was a slippery slope in determining just how far de-extinction could, or should, go. "By the way, if we could 'de-extinct-ify' a mammoth, we could 'de-extinct-ify' a Neanderthal," commented a paleobiologist. "And then you could multiply the ethical issues by a matter of ten-fold" (Interviewee 3).

Far more controversial than practitioners' attempts to resurrect a thylacine or a mammoth were other conjectures that scientists could, in theory, one day resurrect a Neanderthal. In 2013, this exact prospect made international news. In "Can Neanderthals Be Brought Back from the Dead?" *Der Spiegel* quoted Church as suggesting it could be technically possible to bring back a Neanderthal.[25] The news went viral. The *MIT Technology Review* wrote, "Wanted: Surrogate for Neanderthal Baby."[26] The London *Daily Mail* announced, "Wanted: 'Adventurous woman' to give birth to Neanderthal man—Harvard professor seeks mother for cloned cave baby."[27] The *Huffington Post* in Canada covered the news too: "Dream of giving birth to a bouncing Neanderthal baby? One of the world's leading geneticists believes he can make it happen. George Church, a professor at Harvard Medical School, told Der Spiegel we have the technology to not only reconstruct the DNA of our long-extinct relative, but actually resurrect the species."[28] Church's colleagues reacted immediately and critically. Svante Pääbo, for example, wrote a piece for the *New York Times* condemning Church's claim as both technically impossible and ethically inappropriate.[29] Another interviewee shared similar sentiments: "George Church and his suggestion that you can clone a Neanderthal is just idiotic. First of all, the technology—and you never want to say *never*—but the technology is nowhere near it. And secondly, why would you want to? Just to see it?" (Interviewee 18). Church, however, felt his message was gravely misinterpreted by media outlets. In response, *Der Spiegel* tried to clear up the confusion by publishing another piece titled "Surrogate Mother (Not Yet) Sought for Neanderthal."[30] Nonetheless, this controversy highlighted the ethics of research responsibility and that scientists have something to say about the bioethics of de-extinction in particular.

There were a number of reasons why practitioners questioned the motivations for resurrecting extinct species. Some wanted to be clear on the scientific purpose, potential, and payoff. "What's the point of doing it?" asked one evolutionary geneticist. "Is it worth it?" (Interviewee 38). "But why the hell are we doing it?" questioned another researcher much more bluntly (Interviewee 24). "There's no scientific question that could possibly benefit us," remarked yet another. "You don't do science unless it's hypothesis-driven," said a paleontologist. "You want to bring back a dinosaur? What are the questions?" (Interviewee 39). Indeed, a number of interviewees in the field of ancient DNA research opposed bringing back extinct creatures because they think these projects prioritize the sensation of de-extinction over the scientific significance of it.

Additionally, other interviewees argued that de-extinction—regardless of the scientific question motivating it—is an inappropriate use of research or resources. They claimed it is morally wrong to resurrect a mammoth or any other animal for entertainment: "Some people clearly just want to have zoos. . . . All they want to do is put a mammoth in a zoo or a wildlife park," said one molecular biologist. "Are you doing that to make money or are you doing that to inspire people to love nature?" (Interviewee 6). Even if inspiring people to care for or conserve nature is the motivation, others think conservation efforts are better invested in preserving the current environment instead of resurrecting what is already lost: "If you're going to spend the money, it is morally wrong to focus on bringing back an animal that is extinct to the exclusion of an animal that isn't extinct but will be extinct." This same scientist said that if "aliens landed and looked around, . . . they'd be pretty surprised to see that we decided to piss the last of our resources on trying to bring back the mammoth" (Interviewee 2).

Struggles to define de-extinction more generally as science or spectacle was a consequence of the *Jurassic Park* narrative and its influence on the search for DNA from fossils. This legacy sometimes made scientists question the motivations behind de-extinction. Indeed, there were interviewees that viewed resurrection as a gimmick to lure media attention. "It's more hype than science," said a molecular archeologist (Interviewee 35). "It's cheap, it's sleazy, it's the Sunday news. And that's very rarely right, in fact, meaning always wrong," explained a senior scientist (Interviewee 32). "The only reason to do this is to get an article in *National Geographic*" (Interviewee 14). "The only reason to do it is to create a kind of splash and get attention. And that's not a good reason to do science" (Interviewee 47). For interviewees, research significance mattered. Consequently, they viewed practitioners whose projects appeared to prioritize publicity with suspicion. Noting the professional and popular shift in attention from bringing back dinosaurs to mammoth revival, one researcher remarked, "I see the reel being replayed now with . . . how we're going to resurrect mammoths. . . . I know this sounds cynical, but they're exploiting the journalists—the media—for attention" (Interviewee 17). Motivation matters, at least according to interviewees, and when it comes to de-extinction, entertainment potential should never supersede the research question behind it.

Ancient DNA practitioners' opposition to de-extinction, as clearly expressed through these derogatory interpretations of it, might seem difficult to reconcile with the generally positive influence that celebrity—specifically as it related to *Jurassic Park*—had on the field. The intense press and public interest that existed around and affected the search for DNA from fossils mattered at each phase of the discipline's development. Even as scientists rejected the conclusions or the implications of the *Jurassic Park* narrative, they drew on the popularity of the book and movie to emphasize the importance of the technical enterprise in which they were engaged. However, as the concept of resurrecting extinct species evolved from more than a mere speculative idea into a viable endeavor thanks to technological innovations, those within and outside the field of ancient DNA research felt a more urgent need to address it. Ancient DNA researchers' opinions, and chiefly their criticisms, of de-extinction echoed the very same worries that Michael Crichton conveyed in his science-fiction scenario of bringing dinosaurs back to life. Ian Malcolm, the fictional mathematician played by Jeff Goldblum in Steven Spielberg's cinematic adaption of *Jurassic Park*, captured these concerns best with his famous line accusing scientists of being so preoccupied with whether they could that they failed to stop to think about whether they should.

NEWSWORTHINESS

In light of the fact that most in the field of ancient DNA research have little to nothing to do with current de-extinction efforts, why does the idea of resurrecting extinct creatures remain so closely linked to it? And why do a select number of ancient DNA researchers align their research with it, especially given the potential for criticism? According to interviewees, the prospect of resurrecting extinct species is an easy and effortless way to engage with the media. Indeed, in the minds of the broader public, *Jurassic Park* is the ultimate illustration of what ancient DNA researchers may one day accomplish. "I would say probably at least 50% of the time whenever I'm talking to anyone in the media they always ask that question: 'Can we bring thylacines or dodos or mammoths or whatever back to life again?'"

(Interviewee 25). Practitioners' ability to travel back in time genetically speaking, coupled with the prospect that extinction might not be forever, was a perfect recipe for newsworthiness. Consequently, the media returned to ancient DNA's connection to de-extinction repeatedly: "The media love ancient DNA. *They love it.* They *absolutely* love it. Usually, the key question is about cloning. They just can't get enough of it" (Interviewee 3). Even students working in the field nearly thirty years after *Jurassic Park*'s debut get this question too. "It's the first question people ask you," remarked a doctoral researcher. "Honestly, I've been asked that so many times" (Interviewee 53). A second student echoed this experience, giving the phenomenon a name— the "*Jurassic Park* Effect" (Interviewee 54).

Over the past few centuries, media reports have tended to focus on science that is timely, novel, and controversial. Dorothy Nelkin, a science communication scholar, explained that while these sources seek to educate, they also strive to entertain, so science and technology more often become "a source of entertainment than of information."[31] In the early 1900s, for example, Edwin Scripps founded the Science Service, the first official forum for science writing in the United States. Edwin Slosson, the first editor of the service, summarized science in the media as being all about "the fastest or the slowest, the hottest or the coldest, the biggest or the smallest, and in any case, the newest thing in the world."[32] Sharon Dunwoody, another science communication scholar, expands on this, explaining: "Science journalism, again in ways typical of other types of journalism, seeks to hang stories on *traditional news pegs,* characteristics of real-world processes that are proven audience attention-getters."[33] In fact, much remains the same today in the world of science reporting. The search for ancient DNA, especially the first or the oldest DNA, and the controversial potential to bring extinct creatures back to life hit headlines for this very reason.

Ancient DNA practitioners were aware of, even accustomed to, the newsworthiness of their research. "It just gives you access to what you might intuitively think is unreachable, unknown, and mysterious," said one evolutionary geneticist (Interviewee 1). A molecular biologist presented this perspective: "No one really wants to read about the peptidoglycan in bacteria cell walls. It might be very important—probably much more important . . . —but . . . your average person is not going to read that. But you can always write a good story about a king or a mammoth or whatever." Consequently, "We always have journalists ringing us and saying . . ., 'I need a story for something. What have you got?'" (Interviewee 6). Another scientist added, "If you're working on particle physics . . . and you . . . try to . . . explain it to the general public a lot of them might just fall asleep or say, 'Why the hell are we funding this?'" On the other hand, "Ancient DNA is a very easy thing to talk to both the media and the general public about. . . . It's an easy sell for the media to talk about. It's an easy sell for scientists who are in that area to talk to the media about" (Interviewee 25). One geneticist put it this way: "People have tastes and it's a flavor they love" (Interviewee 44).

Ancient DNA's newsworthiness was not just advantageous to media reporters. Its practitioners benefited too. One senior researcher explained, "You've got to separate the ancient DNA research's need for the press, and the press's need for ancient DNA research." According to this interviewee, "The press *loves* ancient DNA because it's often on stories that are very attractive to the general public. . . . Whenever they've got nothing, they come to us because they know it'll be something interesting, right? We work on history. We work on anthropology and archeology. We work on weird shit, dinosaurs, whatever. But the ancient DNA researchers, . . . that's how they justify getting their money, right?" As this interviewee further explained, there are a number of ancient DNA researchers who often take center stage in the public spotlight. Eske Willerslev, a frontrunner in the field, and among the public thanks to the work coming out of his lab, offered an example of this. "That's how Eske gets his money because the Danish government wants to show that Danish science is world class. How better to do that [than] to have *Science* report it, or *National Geographic* or *Discovery Channel* or *Scientific American*." For this scientist, the process of science and science communication is part of one big complex system. "So, Eske gets that press, the government is happy, give Eske more money" (Interviewee 6). Within this system, professional scientific research journals had an important, if not primary, role to play.

For practitioners across every discipline, publishing in top-tier research journals such as *Nature, Science,* and *PNAS* is both a mark of professional prestige and a guaranteed in with the media to publicize their research and its impact. Indeed, the potential to simultaneously achieve both professional and popular renown is a strong draw for many. One interviewee, for example, shared this story to make the point: "I was collaborating on a *really* nice ancient DNA project—good project, good results. They said, 'We're sending this to *Nature*.' I said, 'It's not a *Nature* paper. You're wasting your time.' I said, 'Send it to such-and-such journal.' 'Nope! No!' It went to *Nature:* rejected. Then they tried, I think, PNAS ('previously submitted to *Nature* and *Science*'): rejected. [Smiles] And eventually it ended up in the journal I'd first recommended. [Laughs]" As this interviewee explained, "The top journals . . . are *almost* the link to the popular media. If you look at *Nature,* it is more than a science journal. . . . Although they do publish high-level science, they also like a damn good story. *They do.* You know, *short* papers with a *punchy* headline." This scientist summarized the situation: "So, I think the attempt to get their work into these top journals, repeatedly (with a lot of success I might add in some cases), to some extent colors people looking for what you might call 'sound-bite-research.' 'Let's sequence that hominid.' You know that's going to be a *Nature* paper if that's got DNA in it" (Interviewee 3). The technical literature had a commercial component, and researchers tried to play to both these expectations accordingly.

Indeed, the process of science and science communication is very much a collective enterprise. As science studies scholar Martina Franzen notes, the prestigious research journals such as *Nature* and *Science,* because of their links to mass media, tend to favor spectacular or surprising results.[34] Both of these scientific journals have historically catered to both professional and popular audiences such that news value has affected, and continues to affect, their publishing process as well as the science reporting process.[35] According to sociologist Peter Weingart, "The link of top journals such as *Science* and *Nature* to the mass media by way of pre-publication press releases and related promotional activities that play to the news values of novelty and sensation has an impact on the communication process."[36] Ancient DNA researchers recognize this too. "On the one hand, it makes it very high-profile," said one senior scientist. "It is presumably very much in the mind of *Nature* and *Science* editors when they are considering to accept a paper or not. How much media attention are they going to get and therefore, how many copies are they going to sell? And what [are] citation indexes of the paper going to be and therefore, where [does] their journal sit? That's the sole motivating—well, not sole motivating factor—but significant motivating factor" (Interviewee 32). News values, as determined by the media and top-tier journals, filter back into scientists' decisions concerning the production and presentation of scientific knowledge. Another evolutionary biologist explained, "If I went off and sequenced genomes of three animals in Australia, add some level of hybridization in the past, it would be interesting to me and interesting to a few evolutionary biologists around the world, but it wouldn't be newsworthy or not media-kind-of-newsworthy." Conversely, "If you do it on modern humans, Neanderthals, and Denisovans—because two of them are extinct, one was meant to be a cave-dwelling thug, and the other one no one even realized existed—then that in itself makes it high profile and therefore, it creates greater interest and therefore, greater funding into that kind of research." In other words, "It's like a self-perpetuating system" (Interviewee 25).

Scientists in the spotlight, such as those involved in ancient DNA research and especially those involved in de-extinction research, are often criticized by colleagues who say their science is heavily influenced by popular interest. "Sometimes the research is compromised by the media," commented a molecular archeologist. For this interviewee, this is because the media plays a role in how scientific research can be conceived, conducted, and communicated: "I think it's a little bit dangerous that somehow the media have a big influence on the direction of research, not necessarily on the results but what is interesting to do because it sells so well. . . . Even with the more intellectual higher-ranking journals like *Science* or *Nature* and so on." However, this system affects all of science: "You are a bit forced, sometimes, to publish data premature. . . . I mean, we are living in a capitalistic system and science is connected to it. It's not completely independent of it. So, we all need the money to do our research and to have our own positions safe and so it will always be compromised in some way" (Interviewee 13). Another researcher remarked on the role of funding in particular: "Funding is a huge issue, and I think, unfortunately, it actually really shapes the research that gets done because a lot of people try to chase the trends to try to capture the funding to then get their research done" (Interviewee 27).

Although this tension between science and the spotlight is not unique to the search for DNA from fossils but shared among the sciences, for those practitioners in this specific area of study, it was a particular and constant challenge. In a field so often featured in the media, publicity could come at a cost. One researcher remarked on the negative and positive effects of public attention: "Amongst colleagues, they said, 'Oh, that is the guy that is going on TV. He can't do proper research.' But in the end, I think, it helped me and other programs helped people quite a lot to explain what they do and in order to get funding because that *is* the relation even if people deny it and say, 'Oh, we only do proper science in the lab.' It is" (Interviewee 14). Another leading practitioner described it as a mixture of jealousy and being perceived by colleagues as a "media whore" and "selling out to get more money." However, this same scientist explained the drive to publicize their work despite collegial criticism: "If I thought that it would have no effect on my possibilities of getting another *Nature* or *Science* publication, at all, I would probably say 'no' to participating in the media." But to be on top time and time again, this scientist said, "I'm kind of forced to do something" (Interviewee 7).

Although such actions were a clear part of the scientific enterprise, various scientists in the field of ancient DNA research, as well as those outside of it, held the view that fame and quality research were difficult to reconcile. The life of Carl Sagan, the famous astronomer and science television star who rose to fame in the 1980s, is an excellent example of this. As highlighted by science studies scholar Michael B. Shermer, "So famous did he become that a 'Sagan Effect' took hold in science, whereby one's popularity and celebrity with the general public were thought to be inversely proportional to the quantity and quality of real science being done."[37] This view of stardom—that the greater one's fame, the lesser one's authenticity—is a well-known trope. Graeme Turner, a media studies scholar, explains it this way: "Indeed, the modern celebrity may claim no special achievements other than the attraction of public attention." To elaborate, Turner draws on the example of Kim Kardashian, an American media personality who is often criticized as someone who is only famous for being famous. Turner continues, "As a result, and as the example of Kim Kardashian might suggest, most media pundits would argue that celebrities in the twenty-first century excited a level of public interest that seems, for one reason or another, disproportionate."[38]

In a paper published in *Genome Biology*, the geneticist Neil Hall argues for the rise of a Sagan-like effect in science today but on a Kardashian-style level. "I am concerned that phenomena similar to that of Kim Kardashian may also exist in the scientific community," Hall claims. "I think it is possible that there are individuals who are famous for being famous." To be clear, Hall does not disparage scientists for seeking the spotlight but suggests that in this age of modern media, there needs to be a way to measure a researcher's professional and public persona: "I don't blame Kim Kardashian or her science equivalents for exploiting their fame, who wouldn't? However, I think it's time that we develop a metric that will clearly indicate if a scientist has an overblown public profile so that we can adjust our expectations of them accordingly." He proposes that this new metric—"The Kardashian Index"—could quantify the public profile of a scientist by comparing their social media followers to their number of article citations. Hall's argument, and the fact that it was published in a research journal, is even further evidence that science is not immune from social and cultural influences like celebrity.[39]

Despite the problems that public-facing scientists encounter, there is still an incentive to seek the spotlight. According to Nelkin, researchers and research institutions have long been well aware of and accustomed to this reality and have taken advantage of it: "Individuals try to attract press attention for a variety of reasons—to influence public views, to attract funds, or to establish their competitive position in 'hot' fields of research." Nelkin argued that in the 1970s, scientists in the new field of recombinant DNA research launched a serious media show to hype their research and to combat criticisms, even fears, about the promise or perils of genetic engineering. Nelkin also suggested that scientists have learned how to position their science, through the press and for the public, in a favorable framework for funding: "Geneticists today, seeking to maintain support for costly research, have become skilled in rhetorical strategies designed to attract the media. They describe the genome as a 'bible,' a 'medical crystal ball,' a 'blueprint of life.' They promise that the Human Genome Project will 'unlock the secrets of life,' allowing the prediction and control of disease."[40] Like these practitioners, ancient DNA researchers faced a similar need to bring in resources, particularly during the early years, and they too turned to the public for help.

Media attention can be used as an epistemic strategy across the sciences to claim legitimacy too. As sociologist Massimiano Bucchi suggests, researchers turn to the public for legitimacy at the moment of making a controversial claim, at times of crisis, when there is competition or a desire for cooperation, and when in need of defining and negotiating boundaries of science.[41] In fact, other scholars have identified various examples of this. Bruce V. Lewenstein's work on the cold fusion controversy in the late 1980s and early 1990s highlights how scientists bypassed conventional research and review norms to advance a controversial claim.[42] The discovery of cold fusion was not only announced in the press; the controversy over its reality and means of replication played out in the press too. Similarly, science studies scholar Angela Cassidy makes a compelling case regarding evolutionary psychology and how popular science provided a unique venue outside the norms of academia for debate across disciplines as researchers tried to claim their expertise in this area of research.[43] Felicity Mellor, another science studies researcher, outlines how a select group of planetary scientists and astronomers actively advocated for the threat of an asteroid colliding with earth in the near future. She argues that they promoted the asteroid impact threat via evidence, narratives of technology-to-the-rescue, and appeals to the media in order to confirm the legitimacy of their concerns as an important scientific issue.[44] Research by science communication scholars Rae Goodell, Declan Fahy, and Jane Gregory further demonstrates the role of the individual in shaping public science and how stardom feeds into the shaping of science itself.[45] Even more closely related to the case of ancient DNA research, Amy Fletcher argues that the way scientists framed the thylacine de-extinction project was part of a "deliberate strategy to court media and public attention."[46] Indeed, the ever closer coupling between science and the media is a conscious, calculated, and growing phenomenon.[47]

SCIENCE THAT SELLS

Nearly three decades after the release of *Jurassic Park*, the search for DNA from fossils is still intimately connected to the idea of resurrecting ancient and extinct species, be it dinosaurs, mammoths, thylacines, quaggas, or even passenger pigeons. This link between the two, the science of ancient DNA research and the prospect of de-extinction, was not established by accident. Far from it, scientists, journal editors, media reporters, popular writers, and movie producers alike made the link and reinforced the association as the disciplinary development of ancient DNA research coincided with the worldwide success of *Jurassic Park*.

Ancient DNA researchers then, as well as now, engage with this rhetoric of resurrection, or at least entertain press and public interest in it, not because it represents their work but because they understand the advantages in doing so when communicating to the public for support. As far as some interviewees are concerned, de-extinction can be seen as a way for scientists to marshal resources for further research. "I think de-extinction has proven to be very popular and interesting to the public," remarked a young leading researcher in the field. "And it's also a lightning rod—maybe lightning rod is an exaggeration—but it is a point in which some investors might want to put some money in, and I think that kind of highlights just how underfunded we've become and scientists are reaching out to alternative funding sources." This interviewee further explained, "It is like *Jurassic Park* playing out in real life, but it's also kind of like a means to an end. So, the real value of the Human Genome Project wasn't really getting the human genome. It was all the technologies that came out of it. It was all of the additional things that developed as a result of having a lightning rod to focus money on to get things done." They further illustrated this point: "So, I think Pleistocene Park is much more realistic than *Jurassic Park,* and people love charismatic megafauna. There is a reason why the panda is the lead organism on the World Wildlife Fund. People love big charismatic megafauna and what is more charismatic than a cute, cute woolly mammoth?" (Interviewee 27). Indeed, news value affects more than science reporting; it also affects how and what science gets pursued and published, playing into this self-perpetuating system in which high-profile publications lead to high-profile press that might lead to further funding.

Since the turn of the twenty-first century, new technologies and techniques—such as the innovation of NGS and CRISPR—have caused a number of ancient DNA researchers to do more than just entertain the idea of reviving extinct species. In fact, some have reconsidered the idea of de-extinction as a legitimate prospect, embracing the effort to bring back extinct creatures in light of technological developments and the possibilities they may afford. According to interviewees in the field of ancient DNA research, there are very credible scientists, as well as not so credible scientists, behind de-extinction efforts who have suggested it is a research reality now more than ever. But today, the speculation around resurrection is less about *Jurassic Park* and much more about scientists' ability to revive other creatures such as the thylacine or passenger pigeon. There is even talk about a real-life Pleistocene Park that could be home to the woolly mammoth.[48] This newfound potential has, of course, encouraged further publicity and heightened the celebrity status of ancient DNA research and its connection to bringing extinct species back to life.

Scholars in science communication studies have suggested that the relationship between scientists and the media, journalists in particular, can be described as a "symbiosis"—"that condition in which diverse entities coexist for mutual benefit."[49] Sharon Dunwoody, for example, specifically suggested the idea that journalists and their sources, including scientists, interact within a "shared culture."[50] In fact, this scientist-journalist interaction occurs most often in research subjects that appeal to popular audiences. As cultural studies scholar Peter Broks argues, "popular science is best seen not as a conduit for messages but as a 'forum' where what is popular meets what is scientific." According to Broks, it is best to think of popular science as a "conceptual space"—"a new model for understanding how the meanings of scientific knowledge are challenged and negotiated."[51] In this shared space, however, scientists face difficulties. While they seek to legitimize their research via the mass media, they are also concerned that publicity may compromise their credibility, and by extension their authority. "In other words," Broks further explains, "the problem lies in the expectation that there can be some measure of control over the meanings of an idea once it is placed in the public domain."[52] Dorothy Nelkin put it in these terms with her observation that "while they want their work to be covered in the press, they are constantly concerned about how it is covered."[53]

In fact, the renewed hype for resurrecting extinct species has prompted ancient DNA practitioners to once again balance their celebrity with their credibility, particularly as many de-extinction efforts are a very public affair. One researcher specifically remarked on this challenge: "When you put out a publication you are in so much control of it. Ultimately, you control your product." However, "when you work with the media that's a huge wild card and you have *no* idea what they're going to do or say" (Interviewee 27). Even an editorial in *Nature* notes the tension between scientists' desire for legitimacy and need for authority: "In principle, there is no reason why science should not be accompanied by highly proactive publicity machines. But in practice, such arrangements introduce conflicting incentives that can all too easily undermine the process of the assessment and communication of science."[54] The ever closer connection between science and media, coupled with advantages of popularizing one's work to the broader public, has been and continues to be both a blessing and a curse to scientists who want to communicate to the public while simultaneously sustaining authority over their message regarding its presentation and interpretation across audiences.

Epilogue

BLAST FROM THE PAST

In the spring of 2020, Mary Schweitzer, a pioneer scientist in molecular paleontology at North Carolina State University, along with Alida Bailleul, a postdoctoral researcher of the Institute of Vertebrate Paleontology and Paleoanthropology at the Chinese Academy of Science, and colleagues published evidence for the exceptional preservation of cellular and molecular material in 75-million-year-old dinosaur bones. The evidence came from not one but two juvenile dinosaurs (*Hypacrosaurus stebingeri*), a species of herbivorous duck-billed dinosaur discovered in the badlands of Montana. Specifically, the team had recovered evidence of cells and even structures that resembled chromosomes and DNA retained from calcified dinosaur cartilage. Their findings were based on a series of immunological and histochemical tests conducted at the Museum of the Rockies and replicated separately at North Carolina State University.[1]

The *Global News* in Canada reported on the findings: "Bingo? Possible Dinosaur 'DNA' Found in 75-Million-Year-Old Fossil."[2] The headline was a clear play on words from Mr. DNA, the cartoon character who narrated the step-by-step process of recreating dinosaurs in the *Jurassic Park* movie. *National Geographic* also covered the story, detailing the nuances of the research results and their implications for bringing dinosaurs back to life. "'We're not doing the *Jurassic Park* thing,'" Bailleul reported to the magazine. Schweitzer, all too aware of the cultural connection between the science of ancient DNA research and its science-fiction counterpart (as well as the contentious history between the two), underscored this point: "I'm not even willing to call it DNA because I'm cautious, and I don't want to overstate the results," Schweitzer explained. "There is something in these cells that is chemically consistent with and responds like DNA."[3] Indeed, Schweitzer, Bailleul, and colleagues were careful with their claim because the consensus among scientists was that although DNA could in fact survive the test of time, it was unlikely to persist for more than a million years and even more unlikely to be preserved intact from the days of the dinosaurs. In fact, the upper limit for DNA preservation was just shy of the million-year-mark.

Less than a decade earlier, researchers had extracted DNA from a 700,000-year-old permafrost-preserved horse bone in Alaska and successfully sequenced its genome, making it the oldest genome to date at that time. Ludovic Orlando—a leading practitioner in the field of ancient DNA research at the Center for GeoGenetics at the University of Copenhagen in Denmark—led the effort. *Nature* published their findings. Their work was most impressive for its technological and conceptual developments, as well as the conclusions about evolutionary history that could be drawn from the data they generated. They were able to demonstrate that DNA could survive intact for nearly a million years and they were able to reconstruct the genome of this ancient creature and further research its evolutionary history. To do so, they compared the ancient horse genome to modern genetic data from an approximately forty-thousand-year-old horse, five different breeds of extant horses, and one of the few Przewalski's horses living today—a species once native to Mongolia in Central Asia and thought to be one of the last lines of wild horses. By comparing DNA sequences, they were able to determine that the 700,000-year-old horse was a common ancestor to all living horses today, both wild and domestic. They also determined that the *Equus* lineage—the lineage that includes extant horses, zebras, and donkeys— actually arose 4–4.5 million years ago.[4]

In reporting on this research, the press specifically highlighted the incredibly old age of the fossil and the genome successfully obtained from it. "700,000-Year-Old Horse Genome Shatters Record for Sequencing of Ancient DNA," announced *Wired*.[5] They also speculated about the implications of reaching farther back into the prehistoric past. The *Guardian* of London heralded the feat with the headline "Prehistoric DNA Sequencing:

Jurassic Park Was Not So Wide of the Mark." The article commented: "It is an extraordinary achievement, one that immediately raises the prospect that scientists might soon create the genomes of creatures that died more than a million years ago, possibly several million years. By that reckoning, Crichton and Spielberg would not seem to be so far out."[6] Although journalists highlighted the genome's ancient age as the major breakthrough, scientists argued that the age was not the purpose of the study. Rather, it was the analysis of the genome data that mattered the most. "I don't think we were really pushy in terms of the record. Of course, *Nature* made *all* the titles about it," recalled one researcher involved in the effort. "In the media interviews we played the card, of course, because it's just an easy thing to do." Nonetheless, "the age is not the goal," explained this interviewee. "It was not the principal motivation" (Interviewee 8). Although the age of the genome was a secondary achievement in comparison to the analysis of it, at least according to practitioners, it was nearly impossible for news sources to downplay it.

The record for the oldest genome was especially exciting because throughout the history of the discipline, much interest in the hunt for DNA from fossils had been propelled by the desire to discover the temporal boundaries of molecular preservation in some of the world's most iconic specimens. In fact, the recovery of over 12 billion base pairs of DNA from a nearly 1-million-year-old horse offered an opportunity to reflect on how far the field of ancient DNA research has come since its birth in the early 1980s when researchers at the University of California, Berkeley, claimed to have sequenced a mere 229 base pairs of DNA from a 140-year-old quagga, an extinct member of the horse family. *Nature* published the latter results as the first evidence that DNA could be preserved and recovered from long-dead specimens. Now thirty years later, researchers were clearly pushing the temporality of DNA preservation to an extreme. However, in the field of ancient DNA research, as evidenced through this history of it, records are made to be broken.

In February 2021, for example, *Scientific American* reported the latest and greatest news from the world of ancient DNA research with the following headline: "Mammoth Genomes Shatter Record for Oldest DNA Sequences."[7] The research—conducted by Tom van der Valk and Love Dalén (the Center for Paleogenetics at Stockholm University in Sweden) and colleagues—was published in *Nature*. In this study, researchers extracted DNA from three mammoth teeth found in northeastern Siberia, each from a different time period ranging from the Early to Middle Pleistocene. From these permafrost-preserved specimens, scientists were able to sequence not just one but two ancient genomes from two of the teeth respectively, along with nearly 60 million base pairs of DNA from the third. The DNA recovered from these samples was estimated to be over 1 million years old and offered surprising findings about the evolution and interbreeding of several mammoth species, including the identification of a formerly unknown lineage.[8] A report from ABC Science News states, "It's not quite Jurassic Park, but scientists have successfully extracted and reconstructed million-year-old DNA from mammoth teeth—and solved a couple of mysteries about the creatures' evolutionary history along the way."[9]

From the 1980s to today, technology has been a crucial component in scientists' ability to successfully obtain DNA from increasingly ancient organisms, and with drastic developments from PCR to NGS, much has changed for the field. However, in some ways, some things remain the same. Throughout its growth, the search for DNA from fossils has evolved under the influence of the press and public gaze, particularly as it coincided with and was pulled into the media spotlight by the book and movie *Jurassic Park* in the 1990s. Bailleul and Schweitzer's recent research on the possible preservation of cellular and molecular material from multimillion-year-old dinosaur bones and its unsurprising framing with the *Jurassic Park* narrative is an example of this. Indeed, the media's reporting of the oldest genomes to date, be it from ancient horses or extinct mammoths, with reference to *Jurassic Park* as the benchmark for progress, is further evidence of this enduring connection between science and science fiction.

CELEBRITY SCIENCE: A CONCEPT

Ancient DNA research has historically walked a fine line between science and science fiction. Today, the *Jurassic Park* franchise—as well as the prospect of resurrecting extinct species—continues to be closely connected to the science that inspired it. At the same time, this book has argued that the science of ancient DNA research was in turn influenced by it. Throughout these chapters, I have showed how a newfound scientific practice grew under the influence of consistent and intense press and public interest. More than that, I examined how ancient DNA researchers adapted to the media spotlight, engaging then disengaging with the attention, cultivating then controlling it in response to new innovations and challenges they encountered. Over the decades, the search for DNA from fossils has developed into a celebrity science.

Celebrity science is a new theoretical concept and analytical framework that has surfaced from my synthesis of professional and popular publications on ancient DNA research and from my analysis of interviews with scientists. From this, I suggest that a celebrity science is a subject of science that exists and evolves under intense public interest and extreme media exposure. The media are essential in the making of a celebrity science in that they consistently seek the science and its practitioners for their news value and appeal to the broader public. They repeatedly craft opportunities for publicity. However, it is just as essential that scientists participate in the process too. In such a science, media presence as well as influence is so substantial that researchers respond, positively and negatively, to the attention. They pragmatically anticipate then create their own occasions for fame or they develop strategies for distancing their work from it. They invent and reinvent their reputations and the reputation of their science according to popular influence. These science-media interactions are more than episodic. A celebrity science is the outcome of prolonged publicity: it is the process and product of consistent science-media interactions around a subject of science over an extended period of time. This results in a relationship between researchers and the media that is driven by the public interest they both seek to obtain then maintain.

Crucially, the term "celebrity science" does not seek to examine or explain celebrity on the individual level by pointing to the interactions that a *single scientist* has with the media. Instead, the intent is to explore how celebrity works on the group level. This involves an investigation of the celebrity that exists around a *subject of science*—the content of the research itself, as well as its conclusions and implications—with attention to how it affects the overall community of researchers working within it. In a celebrity science, it is the subject of science that is marketed as a commodity.

Although the term "celebrity science" applies specifically to the group level, I chose it as an extension of a different but not unrelated term—"celebrity scientist"—that does indeed refer to the phenomenon of celebrity at the individual level. This latter term was initially introduced by communication studies scholar Declan Fahy to explain the rising renown and influence of a handful of scientists today. Although the term is relatively recent and specific to the modern movement of celebrity culture in the mid- to late twentieth century, it feels familiar because scientists have in fact had a public presence in the past.[10] Thomas Edison, for example, was the late nineteenth-century image of an iconic inventor, while Albert Einstein was the personification of early twentieth-century physics.[11] Fred Hoyle was the voice of astronomy on the radio, while Carl Sagan was the face of cosmology thanks to television and the on-screen stardom it fostered.[12] Isaac Newton as a popularizer of mathematics is another example of this pattern as early as the seventeenth century.[13]

The rise of mass media in the mid- to late twentieth century and journalists' increased interest in science reporting, however, have offered researchers opportunities to become public-facing practitioners on a new level.[14] In the 1970s, the science communication scholar Rae Goodell highlighted the effects of this, profiling a range of researchers from the anthropologist Margaret Mead and the biologist Paul Ehrlich to the chemist Linus Pauling. According to Goodell, these were all "visible scientists." These visible scientists shared personal and professional characteristics—media-oriented characteristics—that helped them attain press and public visibility. As she aptly argued, visible scientists use their authority and access to the media as a platform from which to speak to the public not just about science but also about science policy.[15] Fahy's celebrity scientist concept builds on this.

For Fahy, the celebrity scientist is a new breed of scientist that has grown out of opportunities afforded by the rise of mass media but more importantly in light of a new celebrity culture. Fahy argues that from the 1980s onward, media started treating scientists as celebrities, and scientists in turn started acting as such. Much of this became apparent at the turn of the century. *The Independent* of London, for example, noted that the science of the early 2000s was full of "media superstars." The *New York Times* called Neil deGrasse Tyson a "space-savvy celebrity," while *Nature* called Susan Greenfield a "celebrity neuroscientist" and *Science* called her a "rock star." As Fahy suggests, there is a qualitative difference between the visible scientists of the past and the celebrity scientists of today, and Carl Sagan embodied that difference.[16] Fahy, in an article written with Bruce Lewenstein, argues that in the 1960s and 1970s, Sagan was what Goodell called a visible scientist. According to these authors, his visibility turned to celebrity after his big television break with his personal but professional show *Cosmos:* "Sagan marked the shift from visible scientist to celebrity scientist; he was a celebrity within a general culture that increasingly valued celebrity for its own sake." This rise of celebrity culture, the increasing value of celebrity for its own sake, was a new social phenomenon introducing a view of "scientific stardom." Fahy profiled a series of scientists who qualified as celebrity scientists, such as cosmologist Stephen Hawking and paleontologist Stephen Jay Gould. They were credentialed experts in their professional spheres but also attained fame, fortune, and influence in the public sphere. As celebrity scientists, they used the media as a public platform to popularize science and influence public attitudes toward it.[17] For Fahy, however, stardom's influence cuts both ways. As celebrity scientists, their stardom affords them influence within science.[18] In other words, stardom filters back into science, affecting the process of science.

According to Fahy and other media studies experts, celebrity and celebrity culture, although observable phenomena in society today, are difficult to define.[19] Indeed, what celebrity is and how it comes to be is far from straightforward. Graeme Turner offers a description of celebrity that does a nice job of capturing its complexity. According to him, celebrity is best understood as both a process and a product: it is both "a genre of representation" and "a commodity traded by the promotions, publicity and media industries that produce these representations and their effects."[20] Celebrity is more than intermittent publicity. The making of celebrity is both a process and a product of repeated science-media interactions.

While celebrity has its positive attributes, such as fame or fortune, it also carries negative connotations. Historically, celebrity has been associated with inauthenticity. The historian Daniel Boorstin, for example, offered a definition that continues to be one of the most well-known adages regarding celebrity today. "The celebrity," according to Boorstin, is "someone who is well-known for their well-knownness."[21] As he further explained, the celebrity is not necessarily known for their achievements but for their ability to publicly differentiate themselves from others via their personality. Boorstin argued that this differentiation is trivial, and he attributes the rise of the celebrity as a consequence of the inauthenticity of contemporary American culture in particular. Celebrity, in this view, represents a struggle between the authentic and inauthentic, and at least in this sense it is a pejorative term. However, as this history of ancient DNA research has revealed, celebrity is a more complex concept than the adage of being well-known for being well-known allows.

Sociologist Chris Rojek, for example, considers celebrity "the attribution of glamorous or notorious status to an individual within the public sphere." For him, glamor and notoriety capture both the favorable and unfavorable forms of public recognition, which are often, and sometimes simultaneously, connected with celebrity. As Rojek argues, celebrity, whatever its attributions and effect, is carefully constructed by the mass media, and it can come in different forms too. Celebrity can be "ascribed" or "achieved," meaning fame can come from lineage (Prince William and Prince Harry, for instance) or from accomplishments (Venus Williams and Serena Williams). However, celebrity can also be "attributed." This happens when media sources repeatedly represent an individual or idea as exceptionally noteworthy.[22] In whatever way stardom manifests itself, the media is a crucial component in the making of celebrity.

In the history of ancient DNA research, celebrity was bestowed on the discipline in more than one way. Celebrity was ascribed through the broader public's inherent interest in fossils and the potential to recover DNA to tell us about ancient history and deep time. At the same time, celebrity was achieved through the accomplishments of scientists in the field sequencing the first or the oldest DNA from some of the world's most charismatic creatures. Celebrity was also attributed by the media, evident from their frequent reporting on the topic over a prolonged period of time.

As the search for DNA from fossils developed into a discipline from the 1980s to today, the researchers involved in it responded to the field's technological challenges and increasing identity as a celebrity science. Practitioners found they had to work around and against two different but not unrelated problems affecting the credibility of the search for DNA from fossils: the problem of contamination as it related to ancient DNA authenticity and the problem of what they viewed to be too much unjustified publicity as a result of the field's celebrity. Scientists were concerned about "contamination" in both a literal and figurative sense.

By the end of the 1990s, contamination concerns as they related to ancient DNA authenticity had placed the practice's credibility on the line. Here, the issue of contamination was illustrated most obviously and most publicly through a number of research papers claiming to have extracted and sequenced multimillion-year-old DNA from various specimens including amber fossils and bone from the days of the dinosaurs. Not long after these papers were published, other practitioners challenged the authenticity of their findings. In fact, some practitioners demonstrated that such bold claims were either irreproducible or the outright product of contamination. The overturning of these research results had devastating consequences for the community's reputation, so much so that researchers within it had to work to reestablish their legitimacy in light of such failed expectations. The ancient DNA community was not just responding to failed expectations, but to the very public nature of those failed expectations because these very studies had been published in high-impact journals like *Nature* and *Science* and broadcast across the mass media. Consequently, scientists were also worried about the influence of celebrity as some saw the hype around the search for DNA from fossils as a further and more figurative source of contamination that had a profound impact on the perceived credibility of the scientific practice.

Interestingly, contamination concerns, despite the division they caused in the community, were also a source of cohesion. In an emerging practice full of multidisciplinary interests and experiences, scientists were united through common problems concerning the preservation, extraction, and sequencing of DNA from fossils. In the late 1980s and early 1990s, contamination was a source of community cohesion as researchers discussed it in newsletters and debated it at conferences. At the turn of the century, it became a very public conflict. But regardless of what side of the schism one fell on, the quest for ancient DNA authenticity was a marked feature of ancient DNA research. Scientists' efforts to "discipline" their discipline via criteria of authenticity ultimately defined the discipline.

Even as contamination was a main source of disciplinary cohesion, so was the celebrity that surrounded the field. In fact, it empowered it, and here the media served as a crucial component to the field's growth in its initial formation and overall identity. Specifically, there were two ways celebrity influenced the discipline's development. First, celebrity influenced the community's formation. This was most evident through the example of *Jurassic Park* and the multimillion-dollar franchise that followed it. The book and movie provided momentum behind the emerging practice, helping to marshal interest in terms of financial and organizational initiatives. This intentional exchange between scientists and the media—specifically around the idea of discovering DNA from some of the world's most ancient and charismatic creatures—influenced publication timing, grant funding, research agendas, and professional recruitment. Persistent publicity on behalf of media reporters—further nurtured by scientists, journal editors, and funding agencies—gave direction to ancient DNA research in its earliest, most vulnerable phase of development.

Next, celebrity played a role in ancient DNA research's identity as a new scientific field. The public recognition gave the field and the practitioners working within it a sense of legitimacy. "Ancient DNA"—with the help of the media—became a brand. This was especially important given the absence of an overarching theoretical framework, as well as in the absence of, or difficulty in obtaining, reliable financial or institutional support. Indeed, this book shows how practitioners' concerns about contamination, as related to both ancient DNA authenticity and the influence of celebrity, and the way scientists articulated those concerns privately as well as publicly, played a fundamental role in driving, even defining, ancient DNA research as a scientific and technological practice in its own right.

CELEBRITY SCIENCE: IN CONTEXT

The history of ancient DNA research offers the opportunity to trace the development of a discipline in a world of modern media and celebrity culture, and at a time when science communication expectations are shifting. By placing the practice in this context, we can begin to understand the how and why behind the making of a celebrity science. In the 1980s, when the field was forming, a number of practitioners in the United Kingdom initiated the Public Understanding of Science (PUS) movement, a systematic endeavor to increase public awareness and appreciation of science and technology.[23] The idea was that improving scientific literacy would encourage public and political support of science and technology. The Royal Society hosted a meeting of practitioners to discuss the objectives of this initiative, which resulted in a report published in 1985 that played a profound part in mobilizing the movement across the United Kingdom, the United States, Europe, and elsewhere.[24] As a result of the report, the Committee on the Public Understanding of Science (COPUS) was created to train scientists to work with journalists and other media outlets, rewarding them for their efforts to engage the public. The report also stimulated the Economic and Social Research Council to fund a research program and journal dedicated to the systematic study of the public understanding of science through surveys and other methods.[25]

In the United States, the American Association for the Advancement of Science took similar steps toward improving public understanding of science and technology with the goal of increasing science literacy, especially science education, across the country. Peter Broks argues that the science communication movement of the 1980s was an attempt to legitimize popularization, giving scientists a professional initiative and incentive to appeal to the public through the media. Broks also suggests that this movement to legitimize popularization was just as much a move to empower science at a time when public support of it (intellectually and financially) was in decline. This decreased support occurred as a consequence of the professionalization of science in the second half of the nineteenth century, with individuals' and institutions' attempts to set themselves apart from the public in order to establish their authority within society. Now, however, to legitimize the social standing of their research, scientists and their scientific institutions once again had to appeal to the public.[26]

Overall, the Public Understanding of Science movement required and rewarded scientists to act as public-facing ones. As science studies scholars Jane Gregory and Steve Miller explain, "In the recent past, many scientists looked at involvement in the popularization of science as something that might damage their career; now, they are being told by the great and the good of science that they have no less than a duty to communicate with the public about their work."[27] However, their ability to communicate on a public platform has also been aided by other developments like the "commercialization" of the mass media and even more recent movements in "mediatization," "medialization," and "celebrification."[28] As some communication scholars have defined it, mediatization simply refers to the increased presence and power of digital media devices in everyday life, whereas the term "medialization" specifically refers to the closer coupling of science and mass media.[29] The phenomenon of "celebrification," on the other hand, references the process by which the mass media makes an individual into a celebrity.[30] These developments, taken together, suggest a setting in which to begin to understand the phenomenon of celebrity science as revealed through the history of ancient DNA research. They also suggest a setting from which to understand how the science communication movement, among other movements, are in turn affecting the process and practice of science.

Researchers working in or around the field of ancient DNA research, for example, recognize the importance of communicating with the public about their work. Specifically, they recognize that the topic of de-extinction is a surefire conversation starter with the media. "There's always going to be some level of celebrity science around trying to recreate extinct species," remarked one researcher (Interviewee 25). This interest in de-extinction and its link to the search for DNA from fossils was not necessarily a link in the daily practice of the science itself. Rather, ancient DNA researchers engage with de-extinction, or at least entertain press and public interest in it, because they understand the advantages of doing so when communicating to the press, and therefore public and political audiences. In reference to the topic of de-extinction, one leading practitioner explained: "There's two ways you can approach that. As a scientist responding to those kinds of inquiries, it's very easy to slam the reporters and say, 'Look. This has absolutely no connection. I don't know why you keep bringing this up.' I don't find that serves my benefit or the benefit of the general public *ever.*" This interviewee further explained that while it is "*very easy* to slam it," it is "also *very easy* to jump on it." According to this scientist, "The question is whether you can have a meaningful discussion in the interstitial spaces between those two extremes" (Interviewee 33). This balance to engage with popular interests while maintaining control over the scientific message communicated to the public is the ever present challenge that ancient DNA researchers encounter.

Many in the field have described their engagement with the media on the topic of de-extinction as a strategy, and one often taken out of necessity. According to these scientists, the whole system and one's level of success in it are tightly coupled with media coverage. An interviewee described this as a "self-perpetuating system" in which high-profile publications lead to high-profile press, which leads to further funding and ongoing research (Interviewee 25). For one practitioner, marketing is key: "It's strategic thinking or writing . . . about how to package science into big picture questions that will get high-profile publications and grant funding. . . . Your ability to sell has actually become more important than your ability to do the science, sadly, given the way that funding has actually gone" (Interviewee 32). As a result, this system has shaped the sort of scientist working in or in close proximity to a celebrity science such as ancient DNA research.

Although ancient DNA researchers have not been profiled specifically as celebrity scientists or visible scientists (although some arguably could be), most are media-savvy scientists in that they have adopted characteristics that make them and their research appealing to the broader public. "I think it's a crowd that's in the limelight," explained one evolutionary biologist. "I think it's a crowd that's used to having to speak to people that are outside of a lot of what they do. And I think it's a crowd that kind of *likes* the idea that it's got a lot of media attention, and therefore, has a lot more practice at it." The interviewee further explained, "I'm sure that there are amazing people studying *Arabidopsis* [a genus of plant] who are equally good communicators, but they don't get the opportunities because who the fuck cares about *Arabidopsis?!*" (Interviewee 22). As such, the overall community of ancient DNA researchers are particularly good communicators. Interviewees also observed that the field has attracted and in fact created researchers who are skilled in identifying celebrity-driven research that would mutually benefit themselves, their institutions, and the press and public they serve. One paleogeneticist remarked, "It's producing a weird type of scientist—I would say a business type of scientist—who kind of knows how to function in this environment." For example, "The people publishing in *Nature* and *Science*, even if it's low hanging fruit, they will be the one more likely to pick up a position at a university because, again, the university wants a researcher who produces media attention" (Interviewee 37). Such researchers often embody ambitious, charismatic, competitive, and media-oriented personalities. Consequently, ancient DNA researchers are not always or simply victims of the next fad. Far from it, they are often shrewd in their ability to adopt current research trends, or even to create them, and adapt when newer options are on the horizon. When the spotlight moves on, these scientists know how to move with it.

From this view, I argue that the emergence of a celebrity science, especially as evidenced through the history of ancient DNA research, has been a positive phenomenon. A celebrity-driven strategy—as clearly and consistently utilized by ancient DNA researchers themselves over the decades— was a valid epistemic approach that practitioners, as well as editors and funders, employed when making choices about research agendas, publication acceptance, and grant funding. A celebrity-driven strategy was productive in generating scientific knowledge that might not have been produced otherwise. Certainly not all research was guided by (or needed to be guided by) its potential to attract popular interest. In fact, ancient DNA researchers sought a celebrity-driven strategy not exclusively but in combination with other data-driven and question-driven approaches. They used a variety of methodologies, sometimes foregrounding celebrity over the questions under study, other times prioritizing the utility of technology over the questions being asked, or focusing on biological or historical questions and using ancient DNA techniques as a means to answering them. These researchers adopted multiple approaches simultaneously and iteratively, assessing the achievability of research results against the accessibility of technology, availability of fossil samples, and the prestige they could gain. To be clear, celebrity was not a solitary factor influencing the discipline's development, but it was a significant one. It helped shape the research practice in terms of influencing the questions scientists asked, the funding they received, and the ways they framed their research when communicating to wider popular and political audiences.

Although I argue that celebrity science is a positive phenomenon, scientists did perceive negative effects related to conducting their research under the media's scrutiny. The field was full of tension as scientists tried to balance the growing celebrity with its credibility, not just once but multiple times throughout its history. Interviewees' disparaging remarks about the celebrity-driven nature of research, and particularly the influence of *Jurassic Park,* do not negate what some scientists saw as positive outcomes from visibility. Rather, it highlights the complexity of celebrity, namely the battle for credibility and the need to balance the two. While practitioners desired to legitimize their research via the mass media, they were also constantly concerned that publicity might compromise their credibility and by extension their authority.[31] Like it or not, this was a reality that scientists had to face, and one they continue to encounter.

In considering the celebrity science concept, it is important to understand that the process of science and technology innovation has never been, and can never be, conceived and conducted outside of the society in which it exists. This history has made this point clear, and countless other scholars have argued and provided much more extensive evidence in support of this worldview. Given that science does not exist outside the influence of society, the emergence of celebrity science is evidence of scientists' hyper-awareness and responsiveness to the very culture in which they and their work are embedded. Interviews with the practitioners themselves provide extremely compelling evidence of the ever closer link between science and media, which stands as a further testament both to the increasing presence and power of the science-media connection in society today and to scientists' pragmatic decisions to adapt to it. These practitioners' perspectives reveal the intensity of the science-media connection and their role in actively participating in it, distancing, or even removing themselves from it, depending on changing circumstances across professional and public landscapes.

The celebrity science concept as told through the disciplinary development of ancient DNA research offers much to the history, philosophy, and sociology of science. The majority of science studies scholarship, for example, has focused on the media's relationship with science by examining a scientist or scientific controversy in the public spotlight in order to understand how the two interact.[32] While the broader science-media connection is evident across contemporary scientific practices, as scholars have well argued, instances of such connections shown to be prominent in shaping the practice of science have been primarily individualistic or episodic. Alternatively, the celebrity science concept argues that the role of the media and practitioners' exchanges with it can be much more expansive. In the search for DNA from fossils, media influence was not limited to a single time, place, event, issue, or individual. Rather, media influence in the field of ancient DNA research was sustained over time through a number of interchanges, to the point that the entirety of the discipline's development from the 1980s onward was shaped, and continues to be shaped, by persistent science-media interactions. The celebrity that surrounded the search for DNA from fossils was central to its birth, growth, and life as a new scientific field. Given this history of ancient DNA research, celebrity is also likely to be a part of the field's future, especially as the science-media connection continues to develop into a more powerful and pervasive phenomenon in the twenty-first century.

NOTES

INTRODUCTION

1. Sergio Bertazzo et al., "Fibres and Cellular Structures Preserved in 75-Million-Year-Old Dinosaur Specimens," *Nature Communications* 6 (June 9, 2015): 7352.

2. Steve Connor, "Scientists Discover Red Blood Cells and Protein from 75-Million-Year-Old Dinosaur Fossils," *The Independent* (London), June 9, 2015, www.indepen dent.co.uk/news/science/scientists-discover-red-blood-and-protein-75-million-year-old-dinosaur-fossils-a32766.html.

3. Raúl J. Cano et al., "Amplification and Sequencing of DNA from a 120–135-Million-Year-Old Weevil," *Nature* 363, no. 6429 (1993): 536–38.

4. Malcolm W. Browne, "DNA from the Age of Dinosaurs Is Found," *New York Times*, June 10, 1993, www.nytimes.com/1993/06/10/us/dna-from-the-age -of-dinosaurs-is-found.html.

5. Tom van der Valk et al., "Million-Year-Old DNA Sheds Light on the Genomic History of Mammoths," *Nature* 591, no. 7849 (2021): 265–69.

6. Russell Higuchi et al., "DNA Sequences from the Quagga, an Extinct Member of the Horse Family," *Nature* 312, no. 5991 (1984): 282–84.

7. See Beth Shapiro and Michael Hofreiter, eds., *Ancient DNA: Methods and Protocols* (New York: Springer, 2012).

8. John R. Tkach, "The Extinct DNA Newsletter," March 1983, Author's Personal Collection (file from John Tkach).

9. For example, "Ancient DNA: The Recovery and Analysis of DNA Sequences from Archaeological Material and Museum Specimens," a conference in Nottingham, England, July 1991, Author's Personal Collection (file from Richard Thomas).

10. Elsbeth Bösl, *Doing Ancient DNA: Zur Wissenschaftsgeschichte der ADNA-Forschung* (Bielefeld, Germany: Verlag, 2017).

11. Martin Jones, *The Molecule Hunt: Archaeology and the Search for Ancient DNA* (New York: Arcade, 2001); Svante Pääbo, *Neanderthal Man: In Search of Lost Genomes* (New York: Basic, 2014); Martin Jones, *Unlocking the Past: How Archaeologists Are Rewriting Human History with Ancient DNA* (New York: Arcade, 2016); David Reich, *Who We Are and How We Got Here: Ancient DNA and the New Science of the Human Past* (New York: Pantheon, 2018).

12. Elsbeth Bösl, "Zur Wissenschaftsgeschichte der ADNA-Forschung," *NTM Zeitschrift für Geschichte der Wissenschaften, Technik und Medizin* 25, no. 1 (2017): 99–142; Bösl, *Doing Ancient DNA*. See also Marianne Sommer, "History in the Gene: Negotiations Between Molecular and Organismal Anthropology," *Journal of the History of Biology* 41, no. 3 (2008): 473–528; Marianne Sommer, *History Within: The Science, Culture, and Politics of Bones, Organisms, and Molecules* (Chicago: University of Chicago Press, 2016); Sarah Abel, "Crossing Disciplinary Lines: Reconciling Social and Genomic Perspectives on the Histories and Legacies of the Transatlantic Trade in Enslaved Africans," *New Genetics and Society* 35, no. 2 (2016): 149–85; and Sarah Abel, "What DNA Can't Tell: Problems with Using Genetic Tests to Determine the Nationality of Migrants," *Anthropology Today* 34, no. 6 (2018): 3–6.

13. My decision to approach the history of ancient DNA research in the context of evolutionary biology presents two challenges. First, the ancient DNA community reflects a wide range of professional influences and interests. Therefore, this book is not an exhaustive account of all the major movements, scientific publications, media articles, or research conclusions that some scientists and scholars might be inclined to include. Instead, I have captured an overall picture of the practice's disciplinary development. Second, the ancient DNA community is a dynamic and diverse one, particularly in the personalities that have contributed to its colorful history. Here, I have represented the community's various viewpoints, including their disagreements, to the best of my ability. In light of these two challenges, it becomes clear that despite practitioners' sometimes steep professional or personal differences, the role played by contamination and celebrity is a common theme running through their memories of their history.

14. See *Oxford English Dictionary*, s.v., "publicity," https://en.oxforddictionaries.com/definition/publicity; s.v., "celebrity," https://en.oxforddictionaries.com/definition/celebrity.

15. Scholars have discussed the role of celebrity in science at the individual level. In the 1970s, science communication scholar Rae Goodell introduced the term "visible scientists." See Goodell, *The Visible Scientists* (Boston: Little, Brown, 1977). In profiling several scientists from the anthropologist Margaret Mead to the astronomer Carl Sagan, Goodell argued that these visible scientists shared personal and professional characteristics (media-oriented characteristics) that helped them attain press and public visibility. They then used this newfound visibility as a platform to speak to the public not just about science but also about science policy. More recently, Declan Fahy has introduced the notion of "celebrity

scientists." See Fahy, *The New Celebrity Scientists: Out of the Lab and into the Limelight* (Lanham, Md.: Rowman and Littlefield, 2015). For Fahy, this is a new type of scientist that has emerged in light of the rise of celebrity culture. These celebrity scientists, like the cosmologist Stephen Hawking and the paleontologist Stephen Jay Gould, were credentialed experts in their professional spheres but also attained fame, fortune, and influence in the public realm. As celebrity scientists, they used the media as a public platform to popularize science and influence public attitudes toward science. According to Fahy, however, stardom's influence cuts both ways, affording them influence outside and within science.

16. My goal is to suggest that the concept of celebrity can be extended from the individual to the group level. Although there are a number of ancient DNA researchers who are internationally well known, I do not focus on how these individuals may or may not qualify as celebrity scientists themselves.

17. For more information, see Simone Rödder, Martina Franzen, and Peter Weingart, eds., *The Sciences' Media Connection—Public Communication and Its Repercussions* (Dordrecht, Netherlands: Springer, 2012).

CHAPTER 1. BEFORE *JURASSIC PARK*

1. Motoko Rich, "Pondering Good Faith in Publishing," *New York Times,* March 8, 2010, www.nytimes.com/2010/03/09/books/09publishers.html.

2. Charles Pellegrino, "Dinosaur Capsule," *Omni* 7 (1985): 38–40, 114–15.

3. Pellegrino, "Dinosaur Capsule," 40; Charles Pellegrino, "Resurrecting Dinosaurs," *Omni* 17 (1995): 68–72.

4. Pellegrino, "Dinosaur Capsule," 114.

5. Pellegrino, "Resurrecting Dinosaurs," 69–70.

6. John Wiley to Charles Pellegrino, March 6, 1986, Author's Personal Collection (file from Charles Pellegrino).

7. For a general overview of the state of knowledge in paleontology at the time, see David M. Raup and Steven M. Stanley, *Principles of Paleontology* (San Francisco: W. H. Freeman, 1971).

8. Philip H. Abelson, "Amino Acids in Fossils," *Science* 119, no. 3096 (1954): 576; Philip H. Abelson, "Paleobiochemistry," *Scientific American* 195 (1956): 83–92; Gordon J. Erdman, Everett M. Marlett, and William E. Hanson, "Survival of Amino Acids in Marine Sediments," *Science* 124, no. 3230 (1956): 1026; Tong-Yun Ho, "The Amino Acid Composition of Bone and Tooth Proteins in Late Pleistocene Mammals," *Proceedings of the National Academy of Sciences of the United States of America* 54, no. 1 (1965): 26–31; E. W. De Jong et al., "Preservation of Antigenic Properties of Macromolecules over 70 Myr," *Nature* 252, no. 5478 (1974): 63–64; Peter Westbroek et al., "Fossil Macromolecules from Cephalopod Shells: Characterization, Immunological Response and Diagenesis," *Paleobiology* 5, no. 2 (1979): 151–67; Stephen Weiner, "Molecular Evolution from the Fossil Record—A Dream or a Reality?" *Paleobiology* 6, no. 1 (1980): 4–5; W. G. Armstrong et al., "Fossil Proteins in Vertebrate Calcified Tissues," *Philosophical Transactions of the Royal Society of London, Series B, Biological Sciences* 301, no. 1106 (1983): 301–43.

9. D. A. Jackson, R. H. Symons, and P. Berg, "Biochemical Method for Inserting New Genetic Information into DNA of Simian Virus 40: Circular SV40 DNA Molecules Containing Lambda Phage Genes and the Galactose Operon of Escherichia Coli," *Proceedings of the National Academy of Sciences of the United States of America* 69, no. 10 (1972): 2904–9; S. N. Cohen et al., "Construction of Biologically Functional Bacterial Plasmids in Vitro," *Proceedings of the National Academy of Sciences of the United States of America* 70, no. 11 (973): 3240–44; Peter E. Lobban and A. A. Kaiser, "Enzymatic End-to-End Joining of DNA Molecules," *Journal of Molecular Biology* 78, no. 3 (1973): 453–71.

10. Frederick Sanger, S. Nicklen, and A. R. Coulson, "DNA Sequencing with Chain-Terminating Inhibitors," *Proceedings of the National Academy of Sciences of the United States of America* 74, no. 12 (1977): 5463–67.

11. "Frederick Sanger—Biographical," Nobel Media AB, 2014, www.nobelprize.org /nobel_prizes/chemistry/laureates/1958/sanger-bio.html.

12. Pellegrino, "Dinosaur Capsule," 40, 114.

13. John Tkach, "A Brief History of the Extinct DNA Study Group," September 1993, Author's Personal Collection (file from John Tkach).

14. Tkach, "A Brief History of the Extinct DNA Study Group," 4.

15. Tkach, "A Brief History of the Extinct DNA Study Group," 4–5.

16. John Tkach, "Evolutionary Immaturity of B-Cell Function as a Possible Cause of the Upper Cretaceous Extinction of Orders Saurischia and Ornithischia," unpublished manuscript, Author's Personal Collection (file from John Tkach).

17. George O. Poinar Jr. and Roberta Poinar, *The Quest for Life in Amber* (Cambridge, Mass.: Perseus, 1994), 64–65.

18. Poinar and Poinar, *The Quest for Life in Amber,* 68–69.

19. George O. Poinar and Roberta Hess, "Ultrastructure of 40-Million-Year-Old Insect Tissue," *Science* 215, no. 4537 (1982): 1241–42.

20. Tkach, "Evolutionary Immaturity of B-Cell Function," 10.

21. John R. Tkach, "The Extinct DNA Newsletter," February 1983, 3, Author's Personal Collection (file from John Tkach).

22. Tkach, "The Extinct DNA Newsletter," February 1983, 1–2.

23. Tkach, "The Extinct DNA Newsletter," March 1983, 4. With reference to the Extinct DNA Study Group's use of the word "paleobiology," it is important to note that the group was not the first to use this terminology. From the 1950s to 1970s, a handful of paleontologists sought to reevaluate and reinvent the discipline of paleontology and its relation to geology, biology, and the modern evolutionary synthesis. Part of the process was a shift in methodology to computational and statistical studies of fossils as a way of rereading the fossil record. By the mid- to late 1970s, these concerted efforts resulted in the creation of a new subdiscipline called "paleobiology." These new "paleobiologists" sought to enhance paleontology's scientific status by contributing to understandings of evolutionary patterns and processes. For more information, see David Sepkoski and Michael Ruse, eds., *The Paleobiological Revolution: Essays on the Growth of Modern Paleontology* (Chicago: University of Chicago Press, 2009); Derek Turner, *Paleontology: A Philosophical Introduction* (Cambridge: Cambridge University

Press, 2011); and David Sepkoski, *Rereading the Fossil Record: The Growth of Paleobiology as an Evolutionary Discipline* (Chicago: University of Chicago Press, 2012).

24. Tkach, "The Extinct DNA Newsletter," March 1983, 8–9.

25. Tkach, "A Brief History of the Extinct DNA Study Group," 13.

26. Poinar and Poinar, *The Quest for Life in Amber*, 92.

27. Tkach, "A Brief History of the Extinct DNA Study Group," 14.

28. Tkach, "The Extinct DNA Newsletter," February 1983, 4.

29. Poinar and Poinar, *The Quest for Life in Amber*, 69, 91.

30. Jurassic Park: The Official Website of Michael Crichton, www.michaelcrichton.com/jurassic-park/; Don Shay and Jody Duncan, *The Making of Jurassic Park: An Adventure 65 Million Years in the Making* (New York: Ballantine, 1993), 3.

31. Robert Bakker, "Dinosaur Renaissance," *Scientific American* 232, no. 4 (1975): 58–79.

32. John H. Ostrom, "Archaeopteryx and the Origin of Flight," *Quarterly Review of Biology* 49, no. 1 (1974): 27–47.

33. Adrian Desmond, *The Hot-Blooded Dinosaurs: A Revolution in Palaeontology* (London: Blond and Briggs, 1975); Robert Bakker, *The Dinosaur Heresies: New Theories Unlocking the Mystery of the Dinosaurs and Their Extinction* (New York: William Morrow, 1986).

34. John R. Horner and Robert Makela, "Nest of Juveniles Provides Evidence of Family Structure Among Dinosaurs," *Nature* 282, no. 5736 (November 1979): 296–98.

35. Luis W. Alvarez et al., "Extraterrestrial Cause for the Cretaceous-Tertiary Extinction," *Science* 208, no. 4448 (1980): 1095–1108; William Glen, ed., *The Mass-Extinction Debates: How Science Works in a Crisis* (Stanford, Calif.: Stanford University Press, 1994).

36. Martin J. S. Rudwick, *The Meaning of Fossils: Episodes in the History of Palaeontology* (Chicago: University of Chicago Press, 1972); Ronald Rainger, *An Agenda for Antiquity: Henry Fairfield Osborn and Vertebrate Paleontology at the American Museum of Natural History, 1890–1935.* (Tuscaloosa: University of Alabama Press, 1991); Peter J. Bowler, *Science for All: The Popularization of Science in Early Twentieth-Century Britain* (Chicago: University of Chicago Press, 2009); Paul D. Brinkman, *The Second Jurassic Dinosaur Rush: Museums and Paleontology in America at the Turn of the Twentieth Century* (Chicago: University of Chicago Press, 2010); Lukas Rieppel, "Bringing Dinosaurs Back to Life: Exhibiting Prehistory at the American Museum of Natural History," *Isis* 103 (2012): 460–90; Chris Manias, "The Lost Worlds of Messmore and Damon: Science, Spectacle, and Prehistoric Monsters in Early-Twentieth Century America," *Endeavour* 40, no. 3 (2016): 163–77; Lukas Rieppel, *Assembling the Dinosaur: Fossil Hunters, Tycoons, and the Making of a Spectacle* (Cambridge, Mass.: Harvard University Press, 2019).

37. Jurassic Park: The Official Website of Michael Crichton; Shay and Duncan, *The Making of Jurassic Park*, 3.

38. Bryan Curtis, "The Cult of 'Jurassic Park,'" *Grantland*, November 7, 2011, http://grantland.com/features/the-cult-jurassic-park/; Bryan Curtis, "3 Nerdy Jurassic Park Footnotes Before You Head Off to See the T. Rex in 3-D," *Grantland*, April 5, 2013, http://grantland.com/hollywood-prospectus/three-nerdy-jurassic-park-footnotes-before-you-head-off-to-see-the-t-rex-in-3-d/.

39. Boyce Rensberger, "Entombed in Amber: Ancient DNA Hints of 'Jurassic Park,'" *Washington Post*, September 25, 1992, www.washingtonpost.com/archive /politics/1992/09/25/entombed-in-amber-ancient-dna-hints-of-jurassic-park /7309d11f-8d62-4589-ba8e-a493392dc6e9/.

40. Poinar and Poinar, *The Quest for Life in Amber*, 153.

41. Jon Turney, *Frankenstein's Footsteps: Science, Genetics, and Popular Culture* (New Haven, Conn.: Yale University Press, 1998); W. J. T. Mitchell, *The Last Dinosaur Book: The Life and Times of a Cultural Icon* (Chicago: University of Chicago Press, 1998).

42. Shay and Duncan, *The Making of* Jurassic Park, 6–8.

43. Pellegrino's dinosaur resurrection hypothesis was also published in two other books shortly after the 1985 *Omni* article. See Charles R. Pellegrino, *Time Gate: Hurtling Backward Through History* (Blue Ridge Summit, Pa.: TAB, 1985), and Charles R. Pellegrino and Jesse A. Stoff, *Darwin's Universe: Origins and Crises in the History of Life* (Blue Ridge Summit, Pa.: TAB, 1986).

44. Michael Crichton, *Jurassic Park* (London: Random Century Group, 1991), acknowledgments.

45. Michael Crichton, *Jurassic Park*, paperback edition (London: Random Century Group, 1991), acknowledgments.

46. John Wiley to Charles Pellegrino, March 6, 1986, Author's Personal Collection (file from Charles Pellegrino).

47. Malcolm W. Browne, "Scientists Study Ancient DNA for Glimpses of Past Worlds," *New York Times*, June 25, 1991, www.nytimes.com/1991/06/25 /science/scientists-study-ancient-dna-for-glimpses-of-past-worlds.html.

48. Charles Pellegrino to Malcolm Browne, June 25, 1991, Author's Personal Collection (file from Charles Pellegrino).

49. Jeffrey M. Duban to George O. Poinar, April 26, 1993, Author's Personal Collection (file from Charles Pellegrino).

50. For information on boundary objects, see Susan Leigh Star and James R. Griesemer, "Institutional Ecology, 'Translations' and Boundary Objects: Amateurs and Professionals in Berkeley's Museum of Vertebrate Zoology, 1907–39," *Social Studies of Science* 19, no. 3 (1989): 387–420; Geoffrey C. Bowker and Susan Leigh Star, *Sorting Things Out: Classifications and Its Consequences* (Cambridge, Mass.: MIT Press, 1999); and Susan Leigh Star, "This Is Not a Boundary Object: Reflections on the Origin of a Concept," *Science, Technology, and Human Values* 35, no. 5 (August 10, 2010): 601–17. See also Elsbeth Bösl, *Doing Ancient DNA: Zur Wissenschaftsgeschichte der ADNA-Forschung* (Bielefeld, Germany: Verlag, 2017). In her work on the history of ancient DNA research, specifically the intersection between genetics and history (i.e., genetic history),

Bösl argues that researchers from different disciplines and scientific backgrounds were united in their interest in old molecules, namely the genetic and
evolutionary information that they could recover from them. She suggests that
this merging of interests around the pursuit of old molecules later resulted in a
transdisciplinary field.

51. For information on the sociology of expectations, see Harro Van Lente and Arie
Rip, "Expectations in Technological Developments: An Example of Prospective
Structures to Be Filled in by Agency," in *Getting New Technologies Together:
Studies in Making Sociotechnical Order,* ed. Cornelis Disco and Barend van der
Meulen (New York: Walter de Gruyter, 1998), 203–9; Harro Van Lente and Arie
Rip, "The Rise of Membrane Technology: From Rhetorics to Social Reality,"
Social Studies of Science 28, no. 2 (1998): 221–54; Nik Brown, Brian Rapport, and
Andrew Webster, eds., *Contested Futures: A Sociology of Prospective Techno-Science*
(Aldershot, U.K.: Ashgate, 2000); Nik Brown and Mike Michael, "A Sociology of
Expectations: Retrospecting Prospects and Prospecting Retrospects," *Technology
Analysis and Strategic Management* 15, no. 1 (2003): 3–18; Nik Brown, "Hope
Against Hype—Accountability in Biopasts, Presents, and Futures," *Science
Studies* 16, no. 2 (2003): 3–21; Mads Borup et al., "The Sociology of Expectations
in Science and Technology," *Technology Analysis and Strategic Management* 18,
nos. 3–4 (2006): 285–98; and Harro van Lente, Charlotte Spitters, and
Alexander Peine, "Comparing Technological Hype Cycles: Towards a Theory,"
Technological Forecasting and Social Change 80 (2013): 1615–28.

CHAPTER 2. IDEAS TO EXPERIMENTS

1. Vincent M. Sarich and Allan C. Wilson, "Immunological Time Scale for
Hominid Evolution.," *Science* 158, no. 3805 (December 1, 1967): 1200–1203;
M. C. King and Allan C. Wilson, "Evolution at Two Levels in Humans and
Chimpanzees," *Science* 188, no. 4184 (1975): 107–16.

2. Sarich and Wilson, "Immunological Time Scale for Hominid Evolution."

3. Émile Zuckerkandl and Linus Pauling, "Molecular Disease, Evolution and
Genetic Heterogeneity," in *Horizons in Biochemistry,* ed. M. Kasha and
B. Pullman (New York: Academic Press, 1962), 189–225.

4. Marianne Sommer, "History in the Gene: Negotiations Between Molecular and
Organismal Anthropology," *Journal of the History of Biology* 41, no. 3 (2008):
473–528; Elsbeth Bösl, "Zur Wissenschaftsgeschichte der ADNA-Forschung,"
NTM Zeitschrift für Geschichte der Wissenschaften, Technik und Medizin 25, no. 1
(2017): 99–142; Michael R. Dietrich, "Paradox and Persuasion: Negotiating the
Place of Molecular Evolution Within Evolutionary Biology," *Journal of the History
of Biology* 31, no. 1 (1998): 85–111.

5. A conversation between Alice Taylor and an unnamed electron microscopist
(presumed but not confirmed to be Roberta Hess) about the preservation of
amber insects was reported in a letter from Taylor to Wilson. In this letter,
Taylor asked Wilson if it would be possible to obtain DNA from insects in amber.
Alice Taylor to Allan Wilson, January 9, 1980, Allan Wilson Papers, series 10,

Research, 1965–1990, reel 47, Bancroft Library, University of California, Berkeley.

6. George O. Poinar Jr. and Roberta Poinar, *The Quest for Life in Amber* (Cambridge, Mass.: Perseus, 1994), 72; Allan Wilson, "Molecular Paleontology: Search for Fossil DNA," National Science Foundation Grant Application, 1984, 12–13, Author's Personal Collection (file from Russell Higuchi).

7. Poinar and Poinar, *The Quest for Life in Amber*, 93–95.

8. Poinar and Poinar, *The Quest for Life in Amber*, 93–95, 73–75; Wilson, "Molecular Paleontology," 12–13.

9. "The Quagga Project," The Quagga Project, 2016, http://quaggaproject.org.

10. Jerold M. Lowenstein, "The Cry of the Quagga," *Pacific Discovery* 384 (1985): 40–42.

11. Lowenstein, "The Cry of the Quagga."

12. Russell Higuchi et al., "DNA Sequences from the Quagga, an Extinct Member of the Horse Family," *Nature* 312, no. 5991 (1984): 282–84.

13. Wilson, "Molecular Paleontology," 4, 2.

14. Wilson, "Molecular Paleontology," 151, 152, 154, 155, 150.

15. Wilson, "Molecular Paleontology," 158.

16. Ronald H. Fritze, *Egyptomania: A History of Fascination, Obsession and Fantasy* (Chicago: University of Chicago Press, 2016).

17. Sarich and Wilson, "Immunological Time Scale for Hominid Evolution"; Thomas J. White and Allan C. Wilson, "Molecular Anthropology," *Evolution* 32, no. 3 (1978): 693–94.

18. In the 1980s, developments in forensic science, like DNA fingerprinting and DNA profiling, emerged in parallel with ancient DNA research. For more information, see Alec J. Jeffreys, Victoria Wilson, and Swee Lay Thein, "Hypervariable 'Minisatellite' Regions in Human DNA," *Nature* 314, no. 6006 (1985): 67–73.

19. Svante Pääbo, *Neanderthal Man: In Search of Lost Genomes* (New York: Basic, 2014), 23–26.

20. Pääbo, *Neanderthal Man*, 24–26.

21. Pääbo, *Neanderthal Man*, 26–28.

22. Pääbo, *Neanderthal Man*, 28–30.

23. Pääbo, *Neanderthal Man*, 26–30.

24. Svante Pääbo, "Uber Den Nachweis von DNA in Altagyptischen Mumien," *Das Altertum* 30 (1984): 213–18; Pääbo, *Neanderthal Man*, 30–32.

25. Svante Pääbo, "Preservation of DNA in Ancient Egyptian Mummies," *Journal of Archaeological Science* 12, no. 6 (1985): 411–17; Pääbo, *Neanderthal Man*, 32–34.

26. Higuchi et al., "DNA Sequences from the Quagga," 284.

27. Pääbo, *Neanderthal Man*, 34.

28. In fact, there were similar studies that were not, but might have been, credited as the first to exhibit evidence of DNA from degraded and damaged material. In 1980, for example, practitioners from Hunan Medical College in China published a paper on the preservation and extraction of DNA from ancient human bodies. See Hunan Medical College, *Study of an Ancient Cadaver in Mawantui Tomb No. 1 of the Han Dynasty in Changsha* (Beijing: Beijing Ancient

Memorial Press, 1980). However, in accounts today, most researchers and reporters credit the paper published on extinct quagga DNA as the first demonstration that nucleic acids could be preserved in and extracted from ancient material. This is likely the case because other published research on the same topic around the same time, such as Pääbo's article on ancient mummy DNA that was published in *Das Altertum* in 1984, were not widely read or recognized by the scientific community.

29. Svante Pääbo, "Molecular Cloning of Ancient Egyptian Mummy DNA," *Nature* 314, no. 6012 (1985): 644–65.

30. Pääbo, *Neanderthal Man*, 35.

31. Alec J. Jeffreys, "Raising the Dead and Buried," *Nature* 312, no. 5991 (1984): 198.

32. "Tissue of Baby Mammoth at Berkeley," *University Bulletin* (Berkeley, Calif.) 26, no. 21 (1978): 110–11. Information about the series of inquiries into a collaboration also comes from an unpublished letter from the president of the National Academy of Sciences of the United States to the vice president of the Academy of Sciences of the Soviet Union. Philip Handler to Yuriy Ovchinnikov, October 7, 1977, Allan Wilson Papers, series 10, Research, 1965–1990, reel 46.

33. "Siberian Baby Mammoth," *New Scientist*, September 1977; "Dima: A Mammoth Undertaking," *Science News* 113, no. 11 (1978): 167; "Russia's Gift: A Well-Aged Mammoth," *San Francisco Examiner*, March 7, 1978; "Tissue of Baby Mammoth at Berkeley"; "UC to Test Slice of Mammoth," *San Francisco Chronicle*, March 8, 1978.

34. Claudine Cohen, *The Fate of the Mammoth: Fossils, Myth, and History* (Chicago: University of Chicago Press, 2002); Ralph O'Connor, *The Earth on Show: Fossils and the Poetics of Popular Science, 1802–1856* (Chicago: University of Chicago Press, 2007).

35. Beth Shapiro, *How to Clone a Mammoth: The Science of De-Extinction* (Princeton, N.J.: Princeton University Press, 2015).

36. Walter Sullivan, "Scientist to Study Mammoth Sample for Clues to Life," *New York Times*, March 9, 1978, www.nytimes.com/1978/03/09/archives/scientist-to-study-mammoth-sample-for-clues-to-life-discovered-last.html.

37. Ellen M. Prager et al., "Mammoth Albumin," *Science* 209, no. 4453 (1980): 287–89.

38. John Noble Wilford, "New Test Links Species over 40,000 Years," *New York Times*, July 11, 1980, www.nytimes.com/1980/07/11/archives/new-test-links-species-over-40000-years-protein-albumin-used.html.

39. Higuchi et al., "DNA Sequences from the Quagga"; Harold M. Schmeck Jr., "Scientists Clone Bits of Genes Taken from Extinct Animal," *New York Times*, June 5, 1984, www.nytimes.com/1984/06/05/science/scientists-clone-bits-of-genes-taken-from-extinct-animal.html. This information is also found in an unpublished grant application from the Allan Wilson Archives: Allan Wilson, "DNA Survival," Biomedical Research Support Grant Application, Allan Wilson Papers, series 10, Research, 1965–1990, reel 48.

40. Diana Ben-Aaron, "Retrobreeding the Woolly Mammoth," *MIT Technology Review*, April 1, 1984, 85.

41. John I. Matill, "Our Shaggy Elephant," *MIT Technology Review,* October 1984, 4; Corey Salsberg, "Resurrecting the Woolly Mammoth: Science, Law, Ethics, Politics, and Religion," *Stanford Technology Law Review* 1 (2000): 1–30.

42. Lewis Clifton, "Mad Scientists Are Cloning Dinosaurs as Weapons of the Future," *National Examiner,* August 7, 1984, 31.

43. "The Resurrection of the Quagga," *New Scientist,* December 13, 1984, 21.

44. Mike Benton, "To Clone a Dinosaur," *New Scientist,* January 17, 1985, 43.

45. Adrian Currie and Kim Sterelny, "In Defence of Story-Telling," *Studies in History and Philosophy of Biological and Biomedical Sciences* 62 (2017): 14–21.

46. Currie and Sterelny, "In Defence of Story-Telling," 16.

CHAPTER 3. TESTING LIMITS

1. Randall K. Saikia et al., "Enzymatic Amplification of β-Globin Genomic Sequences and Restriction Site Analysis for Diagnosis of Sickle Cell Anemia," *Science* 230, no. 4732 (1985): 1350–54; K. Mullis et al., "Specific Enzymatic Amplification of DNA in Vitro: The Polymerase Chain Reaction," *Cold Spring Harbor Symposia on Quantitative Biology* 51 (1986): 263–73; Kary B. Mullis and Fred A. Faloona, "Specific Synthesis of DNA in Vitro via a Polymerase-Catalyzed Chain Reaction," *Methods in Enzymology* 155 (1987): 335–50.

2. "Frederick Sanger—Biographical," Nobel Media AB, 2014, www.nobelprize.org /nobel_prizes/chemistry/laureates/1958/sanger-bio.html.

3. The conceptual, technological, and financial development of PCR is a complex history of interactions among scientists, researchers, and entrepreneurs. See Paul Rabinow, *Making PCR: A Story of Biotechnology* (Chicago: Chicago Unviersity Press, 1996).

4. Jeremy Cherfas, "Genes Unlimited," *New Scientist,* April 19, 1990, 29–33.

5. Svante Pääbo, "Ancient DNA: Extraction, Characterization, Molecular Cloning, and Enzymatic Amplification," *Proceedings of the National Academy of Sciences of the United States of America* 86, no. 6 (1989): 1939–43.

6. Pääbo, "Ancient DNA: Extraction, Characterization, Molecular Cloning, and Enzymatic Amplification," 1943.

7. Svante Pääbo, Russell G. Higuchi, and Allan C. Wilson, "Ancient DNA and the Polymerase Chain Reaction," *Journal of Biological Chemistry* 264, no. 17 (1989): 9709, 9712.

8. Pääbo, Higuchi, and Wilson, "Ancient DNA and the Polymerase Chain Reaction," 9711–12.

9. Robert Paddle, *The Last Tasmanian Tiger: The History and Extinction of the Thylacine* (Cambridge: Cambridge University Press, 2000).

10. Richard H. Thomas et al., "DNA Phylogeny of the Extinct Marsupial Wolf," *Nature* 340, no. 6233 (1989): 465–67.

11. Jerold M. Lowenstein, Vincent M. Sarich, and Barry J. Richardson, "Albumin Systematics of the Extinct Mammoth and Tasmanian Wolf," *Nature* 291, no. 5814 (1981): 409–11.

12. Thomas et al., "DNA Phylogeny of the Extinct Marsupial Wolf," 467.

13. Margaret A. Hughes and David S. Jones, "Body in the Bog but No DNA," *Nature* 323, no. 6085 (1986): 208; Glen H. Doran et al., "Anatomical, Cellular and Molecular Analysis of 8,000-Yr-Old Human Brain Tissue," *Nature* 323, no. 6091 (1986): 803–6.

14. Geoffrey Eglinton, "Marking the Conclusion of the Natural Environment Research Council Special Topic in Biomolecular Palaeontology," Lyell Meeting Volume (Earth Science Directorate, March 1994), Author's Personal Collection (file from Terry Brown).

15. Erika Hagelberg, Bryan Sykes, and Robert Hedges, "Ancient Bone DNA Amplified," *Nature* 342 (1989): 485.

16. "Natural Environment Research Council Special Topic in Biomolecular Palaeontology Community Meeting Programme," Glasgow, Scotland, 1990, Author's Personal Collection (file from Richard Thomas).

17. Edward M. Golenberg et al., "Chloroplast DNA Sequence from a Miocene Magnolia Species," *Nature* 344, no. 6267 (1990): 656–58.

18. William Booth, "Ancient Magnolia Leaf Yields Strands of DNA," *Washington Post,* April 12, 1990, www.washingtonpost.com/archive/politics/1990/04/12/ancient-magnolia-leaf-yields-strands-of-dna/b454fb51-d2bd-4da7-b204-1754581ed1f9.

19. "Genetic Code Found in 17-Million-Year-Old Leaf," *New York Times,* April 12, 1990, www.nytimes.com/1990/04/12/us/genetic-code-found-in-17-million-year-old-leaf.html.

20. Julie Johnson, "The Oldest DNA in the World," *New Scientist,* May 11, 1990, www.newscientist.com/article/mg13017685-300-the-oldest-dna-in-the-world-the-discovery-of-geneticmaterial-that-may-be-16-million-years-old-has-left-molecularpalaeontologists-with-more-questions-than-answers.

21. Svante Pääbo and Allan C. Wilson, "Miocene DNA Sequences—A Dream Come True?" *Current Biology* 1, no. 1 (February 1991): 45–46.

22. Arend Sidow, Allan C. Wilson, and Svante Pääbo, "Bacterial DNA in Clarkia Fossils," *Philosophical Transactions of the Royal Society of London, Series B, Biological Sciences* 333, no. 1268 (1991): 429–33.

23. "Biomolecular Palaeontology Discussion Meeting," agenda, Royal Society, London, 1991, Author's Personal Collection (file from Terry Brown).

24. Martin Jones, *The Molecule Hunt: Archaeology and the Search for Ancient DNA* (New York: Arcade, 2001), 25.

25. Jones, *The Molecule Hunt,* 24, 25.

26. "Ancient DNA: The Recovery and Analysis of DNA Sequences from Archaeological Material and Museum Specimens," conference at the University of Nottingham, England, July 1991, Author's Personal Collection (file from Richard Thomas).

27. "Ancient DNA: The Recovery and Analysis of DNA Sequences from Archaeological Material and Museum Specimens." This information is also from an unpublished document of meeting attendants provided to the author by

Richard Thomas. Thomas, "Ancient DNA Meeting Attendants," July 1991, Author's Personal Collection (file from Richard Thomas).

28. Michael Crichton, *Jurassic Park* (New York: Knopf, 1990).

29. Don Shay and Jody Duncan, *The Making of Jurassic Park: An Adventure 65 Million Years in the Making* (New York: Ballantine, 1993), 6–7.

30. Sharon Begley, "Here Come the DNAsaurs," *Newsweek*, June 14, 1993, 57–59.

31. Allan Wilson, "Molecular Paleontology: Search for Fossil DNA," National Science Foundation Grant Application, 1984, 12–13, Author's Personal Collection (file from Russell Higuchi); George O. Poinar Jr., Hendrik N. Poinar, and Raúl J. Cano, "DNA from Amber Inclusions," in *Ancient DNA: Recovery and Analysis of Genetic Material from Paleontological, Archaeological, Museum, Medical, and Forensic Specimens*, ed. Bernd Herrmann and Susanne Hummel (New York: Springer-Verlag, 1994), 92–103; George O. Poinar Jr. and Roberta Poinar, *The Quest for Life in Amber* (Cambridge, Mass.: Perseus, 1994), 73–75.

32. Crichton, *Jurassic Park* (London edition), 68.

33. Browne, "Scientists Study Ancient DNA for Glimpses of Past Worlds."

34. Jeremy Cherfas, "Ancient DNA: Still Busy After Death," *Science* 253, no. 5026 (1991): 1345, 1356.

35. Cherfas, "Ancient DNA: Still Busy After Death," 1354.

36. Robert Wayne to "Friends of Ancient DNA," Zoological Society of London, 1991, Author's Personal Collection (file from Richard Thomas and Anne Stone); Robert Wayne and Alan Cooper, eds., *Ancient DNA Newsletter* 1, no. 1 (April 1992): 1–43, Author's Personal Collection (files from Richard Thomas and Terry Brown); Robert Wayne and Alan Cooper, eds., *Ancient DNA Newsletter* 1, no. 2 (December 1992): 1–41, Author's Personal Collection (files from Richard Thomas and Terry Brown); Robert Wayne and Alan Cooper, eds., *Ancient DNA Newsletter* 2, no. 1 (February 1994): 1–45, Author's Personal Collection (files from Richard Thomas and Terry Brown).

37. Wayne, "Friends of Ancient DNA," 1.

38. Wayne and Cooper, eds., *Ancient DNA Newsletter*, April 1992, 6–8, 2.

39. Wayne and Cooper, eds., *Ancient DNA Newsletter*, December 1992, 43.

40. Cherfas, "Ancient DNA: Still Busy After Death," 1356.

41. Browne, "Scientists Study Ancient DNA for Glimpses of Past Worlds."

42. Scholars have explored the process of disciplinary development in the sciences with regards to differences between discipline formation and professionalization. See Robert E. Kohler, *From Medical Chemistry to Biochemistry: The Making of a Biomedical Discipline* (Cambridge: Cambridge University Press, 1982); Mary Jo Nye, *From Chemical Philosophy to Theoretical Chemistry: Dynamics of Matter and Dynamics of Disciplines, 1800–1950* (Berkeley: University of California Press, 1993); Lynn Nyhart, *Biology Takes Form: Animal Morphology and the German Universities, 1800–1900* (Chicago: University of Chicago Press, 1995); Vassiliki Betty Smocovitis, *Unifying Biology: The Evolutionary Synthesis and Evolutionary Biology* (Princeton, N.J.: Princeton University Press, 1996); Paul Farber, *Discovering Birds: The Emergence of Ornithology as a Scientific Discipline, 1760–1850* (Baltimore: Johns Hopkins University Press, 1997); and Mark Barrow, *A Passion for Birds: American*

Ornithology After Audubon (Princeton, N.J.: Princeton University Press, 1998). Also see Nathan Reingold, "Definitions and Speculations: The Professionalization of Science in America in the Nineteenth Century," in *The Pursuit of Knowledge in the Early American Republic*, ed. Alexandra Oleson and Sanborn C. Brown (Baltimore: Johns Hopkins University Press, 1976), 33–69; Elizabeth B. Keeney, *The Botanizers: Amateur Scientists in Nineteenth-Century America* (Chapel Hill: University of North Carolina Press, 1992); and Paul Lucier, "The Professional and the Scientist in Nineteenth-Century America," *Isis* 100, no. 4 (2009): 699–732.

43. Simon Schaffer, "Natural Philosophy and Public Spectacle in the Eighteenth Century," *History of Science* 21, no. 1 (1983): 1–43; Jan Golinski, "A Noble Spectacle: Phosphorus and the Public Cultures of Science in the Early Royal Society," *Isis* 80, no. 1 (1989): 11–39; Jan Golinski, *Science as Public Culture: Chemistry and Enlightenment in Britain, 1760–1820* (Cambridge: Cambridge University Press, 1992); Simon Werrett, "Watching the Fireworks: Early Modern Observation of Natural and Artificial Spectacles," *Science in Context* 24, no. 2 (2011): 167–82; Chris Manias, "The Lost Worlds of Messmore and Damon: Science, Spectacle, and Prehistoric Monsters in Early-Twentieth Century America," *Endeavour* 40, no. 3 (2016): 163–77; Amy Fletcher, "Digging Up the Past: Paleogenomics as Science and Spectacle," APSA 2009 Toronto Meeting Paper, https://papers.ssrn.com/sol3/papers.cfm?abstract_id=1451865; Amy Fletcher, "Genuine Fakes: Cloning Extinct Species as Science and Spectacle," *Politics and the Life Sciences* 29, no. 1 (2010): 48–60; Jon Agar, *Science and Spectacle: The Work of Jodrell Bank in Post-War British Culture* (Amsterdam: Harwood Academic, 1998).

CHAPTER 4. DINOSAUR DNA

1. Rob DeSalle et al., "DNA Sequences from a Fossil Termite in Oligo-Miocene Amber and Their Phylogenetic Implications," *Science* 257, no. 5078 (1992): 1933–36.

2. Malcolm W. Browne, "40-Million-Year-Old Extinct Bee Yields Oldest Genetic Material," *New York Times*, September 25, 1992, www.nytimes.com/1992/09/25/us/40-million-year-old-extinct-bee-yields-oldest-genetic-material.html; Boyce Rensberger, "Entombed in Amber: Ancient DNA Hints of 'Jurassic Park,'" *Washington Post*, September 25, 1992, www.washingtonpost.com/archive/politics/1992/09/25/entombed-in-amber-ancient-dna-hints-of-jurassic-park/7309d11f-8d62-4589-ba8e-a493392dc6e9; Kathryn Hoppe, "Brushing the Dust Off Ancient DNA," *Science News*, October 24, 1992, 280–81; Virginia Morell, "30-Million-Year-Old DNA Boosts an Emerging Field," *Science* 257, no. 5078 (1992): 1860–62.

3. "Amber: Window to the Past," American Museum of Natural History, 1996, http://lbry-web-007.amnh.org/digital/index.php/items/show/39273.

4. David A. Grimaldi, *Amber: Window to the Past* (New York: Harry N. Abrams, 1996); David A. Grimaldi, "Captured in Amber," *Scientific American*, April 1996, 70–77.

5. Raúl J. Cano, Hendrik N. Poinar, and George O. Poinar Jr., "Isolation and Partial Characterisation of DNA from the Bee Proplebeia Dominicana (Apidae: Hymenoptera) in 25–40 Million Year Old," *Medical Science Research* 20, no. 7 (1992): 249–51.

6. Morell, "30-Million-Year-Old DNA Boosts an Emerging Field," 1860.

7. Rensberger, "Entombed in Amber."

8. Rensberger, "Entombed in Amber."

9. Morell, "30-Million-Year-Old DNA Boosts an Emerging Field," 1861.

10. Browne, "40-Million-Year-Old Extinct Bee Yields Oldest Genetic Material."

11. Raúl J. Cano et al., "Amplification and Sequencing of DNA from a 120–135-Million-Year-Old Weevil," *Nature* 363, no. 6429 (1993): 536–38; "Jurassic Park (1993)," IMDb, www.imdb.com/title/tt0107290/; David A. Kirby, *Lab Coats in Hollywood: Science, Scientists, and Cinema* (Cambridge, Mass.: MIT Press, 2013).

12. Malcolm W. Browne, "DNA from the Age of Dinosaurs Is Found," *New York Times,* June 10, 1993, www.nytimes.com/1993/06/10/us/dna-from-the-age-of-dinosaurs-is-found.html.

13. George O. Poinar Jr. and Roberta Poinar, *The Quest for Life in Amber* (Cambridge, Mass.: Perseus, 1994), 154.

14. Stephen Jay Gould, "Dinomania," in *Dinosaur in a Haystack: Reflections in Natural History* (London: Jonathan Cape, 1996), 225–26; Kirby, *Lab Coats in Hollywood,* 139.

15. Pat H. Broeske, "Promoting 'Jurassic Park,'" *Entertainment,* March 12, 1993, http://ew.com/article/1993/03/12/promoting-jurassic-park/.

16. "Jurassic Park (1993)," Box Office Mojo, www.boxofficemojo.com/movies/?page=daily&id=jurassicpark.htm.

17. "Jurassic Park (1993) Awards," IMDb, www.imdb.com/title/tt0107290/awards.

18. Michele Pierson, "CGI Effects in Hollywood Science-Fiction Cinema, 1989–95: The Wonder Years," *Screen* 40, no. 2 (1999): 158–76; Michele Pierson, *Special Effects: Still in Search of Wonder* (New York: Columbia University Press, 2002).

19. Julia Hallam and Margaret Marshment, *Realism and Popular Cinema* (Manchester, U.K.: Manchester University Press, 2000); Joel Black, *The Reality Effect* (New York: Routledge, 2002); Sheldon Hall and Steve Neale, *Epics, Spectacles, and Blockbusters: A Hollywood History* (Detroit: Wayne State University Press, 2010); Kirby, *Lab Coats in Hollywood.*

20. Pierson, "CGI Effects in Hollywood Science-Fiction Cinema," 166, 167.

21. Dennis McLellan, "Michael Crichton Dies at 66; Bestselling Author of 'Jurassic Park' and Other Thrillers," *Los Angeles Times,* November 6, 2008, www.latimes.com/local/obituaries/la-me-crichton6-2008nov06-story.html.

22. Pamela McClintock, "Steven Spielberg's Top 10 Box Office Successes," Hollywood Reporter, 2015, www.hollywoodreporter.com/news/steven-spielberg-s-top-10-803126.

23. Sharon Begley, "Here Come the DNAsaurs," *Newsweek,* June 14, 1993, 57.

24. Peter H. King, "'Step Right Up and See the Science,'" *Los Angeles Times*, June 16, 1993, http://articles.latimes.com/1993-06-16/news/mn-3654_1_dna-research; Kirby, *Lab Coats in Hollywood*.

25. King, "'Step Right Up and See the Science.'"

26. David Kirby highlights these and other similar interactions and their implications for understanding the relationship between science and media, specifically how science influences, or is in turn influenced by, Hollywood and the blockbuster phenomenon. See David A. Kirby, "Science Consultants, Fictional Films, and Scientific Practice," *Social Studies of Science* 33, no. 2 (2003): 231–68; David A. Kirby, "Scientists on the Set: Science Consultants and the Communication of Science in Visual Fiction," *Public Understanding of Science* 12 (2003): 261–78; Kirby, *Lab Coats in Hollywood*; and David A. Kirby, "Science and Technology in Film: Themes and Representations," in *Routledge Handbook of Public Communication of Science and Technology*, 2nd edition, ed. Massimiano Bucchi and Brian Trench (London: Routledge, 2014), 97–112.

27. Susan Gallagher, "Maverick Dinosaur Expert Gets in His Digs in Montana," *Los Angeles Times*, November 21, 1993, www.latimes.com/archives/la-xpm-1993-11-21-mn-59211-story.html.

28. Virginia Morell, "Dino DNA: The Hunt and the Hype," *Science* 261, no. 5118 (1993): 160.

29. John R. Horner and Ernst Vyse, "An Attempt to Extract DNA from the Cretaceous Dinosaur *Tyrannosaurus rex*," National Science Foundation, 1993, www.nsf.gov/awardsearch/showAward?AWD_ID=9311542.

30. Ben Macintyre, "Fossil Find Brings *Jurassic Park* Closer," *The Times* (London), July 2, 1993, 16; Kirby, *Lab Coats in Hollywood*, 139.

31. Malcolm W. Browne, "Cells of Dinosaurs Apparently Found," *New York Times*, July 1, 1993, www.nytimes.com/1993/07/01/us/cells-of-dinosaur-apparently-found.html.

32. Morell, "Dino DNA," 161.

33. Morell, "Dino DNA," 160.

34. Gerard Muyzer et al., "Preservation of the Bone Protein Osteocalcin in Dinosaurs," *Geology* 20 (1992): 871–74.

35. L. R. Gurley et al., "Proteins in the Fossil Bone of the Dinosaur, Seismosaurus," *Journal of Protein Chemistry* 10, no. 1 (1991): 75–90.

36. Morell, "Dino DNA," 160, 161.

37. Morell, "30-Million-Year-Old DNA Boosts an Emerging Field," 1860–62.

38. Morell, "Dino DNA," 160.

39. "Ancient DNA: Second International Conference," Washington, D.C., October 1993, Author's Personal Collection (file from Richard Thomas).

40. Joshua Fischman, "Going for the Old: Ancient DNA Draws a Crowd," *Science* 262, no. 5134 (1993): 655.

41. Terence Brown and Keri Brown, "Ancient DNA and the Archaeologist," *Antiquity* 66 (1992): 10–23; Terence A. Brown et al., "Biomolecular Archaeology of Wheat: Past, Present and Future," *Biomolecular Archaeology* 25, no. 1 (1993): 64–73; W. Kelley Thomas et al., "Spatial and Temporal Continuity of Kangaroo

Rat Populations Shown by Sequencing Mitochondrial DNA from Museum Specimens," *Journal of Molecular Evolution* 31 (1990): 101–12; Carey Krajewski et al., "Phylogenetic Relationships of the Thylacine (Mammalia: Thylacinidae) Among Dasyuroid Marsupials: Evidence from Cytochrome b DNA Sequences," *Proceedings of the Royal Society, Series B, Biological Sciences* 250, no. 1327 (1992): 19–27; Alan Cooper et al., "Independent Origins of New Zealand Moas and Kiwis," *Proceedings of the National Academy of Sciences of the United States of America* 89, no. 18 (1992): 8741–44; Catherine Hänni et al., "Amplification of Mitochondrial DNA Fragments from Ancient Human Teeth and Bones," *Comptes Rendus de l'Academie Des Sciences, Serie III, Sciences de La Vie* 310, no. 9 (1990): 365–70; Robert K. Wayne and S. M. Jenks, "Mitochondrial DNA Analysis Implying Extensive Hybridization of the Endangered Red Wolf Canis Rufus," *Nature* 351, no. 6327 (1991): 565–68; Susanne Hummel and Bernd Herrmann, "Y-Chromosome-Specific DNA Amplified in Ancient Human Bone," *Naturwissenschaften* 78 (1991): 266–67; Matthias Höss et al., "Excrement Analysis by PCR," *Nature* 359 (1992): 199; Erika Hagelberg and John B. Clegg, "Genetic Polymorphisms in Prehistoric Pacific Islanders Determined by Analysis of Ancient Bone DNA," *Proceedings of the Royal Society, Series B, Biological Sciences* 252, no. 1334 (1993): 163–70.
42. Scott R. Woodward, Nathan J. Weyand, and Mark Bunnell, "DNA Sequence from Cretaceous Period Bone Fragments," *Science* 266, no. 5188 (1994): 1230.
43. "Dinosaur DNA," *New Scientist,* November 26, 1994, www.newscientist.com /article/mg14419532-000-dinosaur-dna/; Robert Lee Hotz, "Bone Yields Dinosaur DNA, Scientists Believe," *Los Angeles Times,* November 18, 1994, www.latimes.com/archives/la-xpm-1994-11-18-mn-64303-story.html; John Noble Wilford, "A Scientist Says He Has Isolated Dinosaur DNA," *New York Times,* November 18, 1994, www.nytimes.com/1994/11/18/us/a-scientist-says-he-has-isolated-dinosaur-dna.html.
44. R. Monastersky, "Dinosaur DNA: Is the Race Finally Over?" *Science News* 146, no. 21 (1994): 324.
45. Amy Fletcher, "Genuine Fakes: Cloning Extinct Species as Science and Spectacle," *Politics and the Life Sciences* 29, no. 1 (2010): 49.
46. Kirby, *Lab Coats in Hollywood,* 227, 228.
47. Kirby, "Science Consultants, Fictional Films, and Scientific Practice"; Kirby, "Scientists on the Set"; Kirby, *Lab Coats in Hollywood.*
48. Kirby, *Lab Coats in Hollywood,* 139, 133–36. Kirby further discusses other cases in which science consultants worked with movie producers to advise on films from *Outbreak* (1995) and *GATTACA* (1997) to *Mission to Mars* (2000) and *Frequency* (2006). See pages 48–49.
49. Kirby, *Lab Coats in Hollywood,* 58.
50. Robert Wayne and Alan Cooper, eds., *Ancient DNA Newsletter* 1, no. 2 (December 1992): 6, Author's Personal Collection (files from Richard Thomas and Terry Brown).
51. Morell, "Dino DNA," 161.

CHAPTER 5. IMPOSING LIMITS

1. Tomas Lindahl, "Instability and Decay of the Primary Structure of DNA," *Nature* 362, no. 6422 (1993): 709–15; Tomas Lindahl, "Recovery of Antediluvian DNA," *Nature* 365, no. 6448 (1993): 700.

2. Edward M. Golenberg et al., "Chloroplast DNA Sequence from a Miocene Magnolia Species," *Nature* 344, no. 6267 (1990): 656–58; Raúl J. Cano, Hendrik N. Poinar, and George O. Poinar Jr., "Isolation and Partial Characterisation of DNA from the Bee Proplebeia Dominicana (Apidae: Hymenoptera) in 25–40 Million Year Old," *Medical Science Research* 20, no. 7 (1992): 249–51; Rob DeSalle et al., "DNA Sequences from a Fossil Termite in Oligo-Miocene Amber and Their Phylogenetic Implications," *Science* 257, no. 5078 (1992): 1933–36; Raúl J. Cano et al., "Amplification and Sequencing of DNA from a 120–135-Million-Year-Old Weevil," *Nature* 363, no. 6429 (June 10, 1993): 536–38; Lindahl, "Recovery of Antediluvian DNA," 700.

3. Lindahl, "Recovery of Antediluvian DNA," 700.

4. George O. Poinar Jr., "Recovery of Antediluvian DNA," *Nature* 365, no. 6448 (1993): 700.

5. Lindahl, "Recovery of Antediluvian DNA," 700.

6. Svante Pääbo, *Neanderthal Man: In Search of Lost Genomes* (New York: Basic, 2014), 58.

7. Svante Pääbo, "Ancient DNA: Extraction, Characterization, Molecular Cloning, and Enzymatic Amplification," *Proceedings of the National Academy of Sciences of the United States of America* 86, no. 6 (1989): 1939–43; Svante Pääbo, Russell G. Higuchi, and Allan C. Wilson, "Ancient DNA and the Polymerase Chain Reaction," *Journal of Biological Chemistry* 264, no. 17 (1989): 9709–12.

8. Pääbo, *Neanderthal Man*, 52.

9. Scott R. Woodward, Nathan J. Weyand, and Mark Bunnell, "DNA Sequence from Cretaceous Period Bone Fragments," *Science* 266, no. 5188 (1994): 1229–32.

10. Robert Lee Hotz, "Bone Yields Dinosaur DNA, Scientists Believe," *Los Angeles Times*, November 18, 1994, www.latimes.com/archives/la-xpm-1994-11-18-mn-64303-story.html; John Noble Wilford, "A Scientist Says He Has Isolated Dinosaur DNA," *New York Times*, November 18, 1994, www.nytimes.com/1994/11/18/us/a-scientist-says-he-has-isolated-dinosaur-dna.html.

11. Woodward, Weyand, and Bunnell, "DNA Sequence from Cretaceous Period Bone Fragments"; S. Blair Hedges and Mary Schweitzer, "Detecting Dinosaur DNA," *Science* 268, no. 5214 (1995): 1191–92; Henikoff Steven, "Detecting Dinosaur DNA," *Science* 268, no. 5214 (1995): 1192; Marc W. Allard, Deshea Young, and Yentram Huyen, "Detecting Dinosaur DNA," *Science* 268, no. 5214 (1995): 1192; H. Zischler et al., "Detecting Dinosaur DNA," *Science* 268, no. 5214 (1995): 1192–93.

12. Ann Gibbons, "Possible Dino DNA Find Is Greeted with Skepticism," *Science* 266, no. 5188 (1994): 1159.

13. Hedges and Schweitzer, "Detecting Dinosaur DNA," 1191.

14. Steven, "Detecting Dinosaur DNA"; Allard, Young, and Huyen, "Detecting Dinosaur DNA"; Zischler et al., "Detecting Dinosaur DNA."

15. Zischler et al., "Detecting Dinosaur DNA."

16. Pääbo, *Neanderthal Man*, 60.

17. Zischler et al., "Detecting Dinosaur DNA, "1193; Jones, *The Molecule Hunt*, 31–38.

18. Richard Monastersky, "Dinosaur DNA Claim Dismissed as a Mistake," *Science News* 248, no. 23 (1995): 373.

19. Peter Aldhous, "'Jurassic DNA' Looks Distinctly Human," *New Scientist* 145, no. 1964 (1995): 5.

20. Malcolm W. Browne, "Critics See Humbler Origin of 'Dinosaur' DNA," *New York Times*, June 20, 1995, www.nytimes.com/1995/06/20/science/critics-see-humbler-origin-of-dinosaur-dna.html.

21. Jeffrey L. Bada et al., "Amino Acid Racemization in Amber-Entombed Insects: Implications for DNA Preservation," *Geochimica et Cosmochimica Acta* 58, no. 14 (1994): 3131–35; Robert F. Service, "Just How Old Is That DNA, Anyway?" *Science* 272, no. 5263 (1996): 810; Hendrik Poinar et al., "Amino Acid Racemization and the Preservation of Ancient DNA," *Science* 272, no. 5263 (1996): 864–66.

22. Poinar et al., "Amino Acid Racemization and the Preservation of Ancient DNA," 865.

23. Poinar et al., "Amino Acid Racemization and the Preservation of Ancient DNA"; Service, "Just How Old Is That DNA, Anyway?"

24. Geoffrey Eglinton, ed., "ABI Newsletter 1," *Ancient Biomolecules Initiative* 1 (November 1995): 1–39, Author's Personal Collection (file from Richard Thomas); Geoffrey Eglinton, Barbara Knowles, and Ursula Edmunds, eds., "Molecular Signatures from the Past," *Ancient Biomolecules Initiative*, Grand Finale (Program and Abstracts), Natural Environment Research Council (1998), Author's Personal Collection (file from Richard Thomas).

25. Eglington, ed., "ABI Newsletter 1," 4–5.

26. "Modern Research into Ancient Biomolecules," *Molecular Biology*, September 1994, 5.

27. Geoffrey Eglinton, "Ancient Biomolecules Initiative Newsletter," *Natural Environment Research Council* 2 (May 1996): 1–2.

28. Eglinton, ed., "ABI Newsletter 1"; Eglinton, Knowles, and Edmunds, eds., "Molecular Signatures from the Past."

29. Andrew B. Smith, "Application for an ABI Research Grant," Natural Environment Research Council, September 1995, 16, 1, Author's Personal Collection (file from Richard Thomas).

30. Smith, "Application for an ABI Research Grant," 3.

31. Jeremy J. Austin et al., "Problems of Reproducibility—Does Geologically Ancient DNA Survive in Amber-Preserved Insects?" *Proceedings of the Royal Society, Series B, Biological Sciences* 264, no. 1381 (1997): 467–74.

32. Smith, "Application for an ABI Research Grant," 15.

33. Austin et al., "Problems of Reproducibility," 470.

34. D. E. Howland and G. M. Hewitt, "DNA Analysis of Extant and Fossil Beetles," in Geoffrey Eglinton, "Marking the Conclusion of the Natural Environment Research Council Special Topic in Biomolecular Palaeontology," Lyell Meeting Volume (Earth Science Directorate, March 1994), 49–51, Author's Personal Collection (file from Terry Brown); J. Pawlowski et al., "Attempted Isolation of DNA from Insects Embedded in Baltic Amber," *Inclusion* 22 (1996): 12–13.

35. Austin et al., "Problems of Reproducibility," 473.

36. Constance Holden, "'No Go' for *Jurassic Park*–Style Dinos," *Science* 276, no. 5311 (1997): 361.

37. Bryan Sykes, "Lights Turning Red on Amber," *Nature* 386 (1997): 764–65.

38. "Ancient DNA III" conference, Oxford, England, July 1995, Author's Personal Collection (file from Richard Thomas).

39. Nigel Williams, "The Trials and Tribulations of Cracking the Prehistoric Code," *Science* 269, no. 5226 (1995): 923.

40. Robert Wayne and Alan Cooper, eds., *Ancient DNA Newsletter* 1, no. 2 (December 1992): 3, Author's Personal Collection (files from Richard Thomas and Terry Brown).

41. Lindahl, "Recovery of Antediluvian DNA," 700.

42. Svante Pääbo, Matthias Höss, and N. K. Vereshchagin, "Mammoth DNA Sequences," *Nature* 370, no. 6488 (1994): 333.

43. Erika Hagelberg et al., "DNA from Ancient Mammoth Bones," *Nature* 370, no. 6488 (1994): 333–34.

44. Alan Cooper et al., "Independent Origins of New Zealand Moas and Kiwis," *Proceedings of the National Academy of Sciences of the United States of America* 89, no. 18 (1992): 8741–44.

45. Matthias Krings et al., "Neandertal DNA Sequences and the Origin of Modern Humans," *Cell* 90 (1997): 19–30.

46. William King, "The Reputed Fossil Man of the Neanderthal," *Quarterly Journal of Science* 1 (1864): 88–97; Ralf W. Schmitz et al., "The Neandertal Type Site Revisited: Interdisciplinary Investigations of Skeletal Remains from the Neander Valley, Germany," *Proceedings of the National Academy of Sciences of the United States of America* 99, no. 20 (2002): 13342–47; Paige Madison, "The Most Brutal of Human Skulls: Measuring and Knowing the First Neanderthal," *British Journal for the History of Science* 49, no. 3 (2016): 411–32.

47. Ann Gibbons, *The First Human: The Race to Discover Our Earliest Ancestors* (New York: Anchor, 2007); Sigrid Schmalzer, *The People's Peking Man: Popular Science and Human Identity in Twentieth-Century China* (Chicago: University of Chicago Press, 2008); Chris Manias, "Sinanthropus in Britain: Human Origins and International Science, 1920–1939," *British Journal for the History of Science* 48, no. 2 (2015): 289–319; Amanda Rees, "Stories of Stones and Bones: Disciplinarity, Narrative and Practice in British Popular Prehistory, 1911–1935," *British Journal for the History of Science* 49, no. 3 (2016): 433–51; Madison, "The Most Brutal of Human Skulls."

48. Chris Stringer, *Lone Survivors: How We Came to Be the Only Humans on Earth* (New York: St. Martin's, 2012).

49. Rebecca L. Cann, Mark Stoneking, and Allan C. Wilson, "Mitochondrial DNA and Human Evolution," *Nature* 325, no. 6099 (1987): 31–36.

50. Harold M. Schmeck Jr., "Intact Genetic Material Extracted from an Ancient Egyptian Mummy," *New York Times*, April 16, 1985, www.nytimes.com/1985 /04/16/science/intact-genetic-material-extracted-from-an-ancient-egyptian- mummy.html.

51. Krings et al., "Neandertal DNA Sequences and the Origin of Modern Humans," 22.

52. Krings et al., "Neandertal DNA Sequences and the Origin of Modern Humans," 19–30.

53. The debate in evolutionary anthropology centered around the origins of human history with evolutionary anthropologists usually subscribing to one of two hypotheses: the Out-of-Africa Model or the Multiregional Continuity Model. The former proposes that humans originated in Africa and then migrated to other parts of the world, while the second suggests that prehumans originated in Africa but then evolved into modern humans after they migrated out of the continent. See Stringer, *Lone Survivors*.

54. Pääbo, *Neanderthal Man*, 18.

55. Patricia Kahn and Ann Gibbons, "DNA from an Extinct Human," *Science* 277, no. 5323 (1997): 176–78.

56. Ryk Ward and Chris Stringer, "A Molecular Handle on the Neanderthals," *Nature* 388, no. 6639 (1997): 225–26.

57. Chris Mihill, "We're African, No Bones About It," *The Guardian* (London), July 11, 1997.

58. Roger Lewin, "Back from the Dead," *New Scientist*, October 18, 1997, 43.

59. Tomas Lindahl, "Facts and Artifacts of Ancient DNA," *Cell* 90, no. 1 (1997): 2.

60. Lewin, "Back from the Dead," 42.

61. Erika Hagelberg, Bryan Sykes, and Robert Hedges, "Ancient Bone DNA Amplified," *Nature* 342 (1989): 485.

62. Erika Hagelberg and John B. Clegg, "Genetic Polymorphisms in Prehistoric Pacific Islanders Determined by Analysis of Ancient Bone DNA," *Proceedings of the Royal Society, Series B, Biological Sciences* 252, no. 1334 (1993): 163–70; Hagelberg, Sykes, and Hedges, "Ancient Bone DNA Amplified."

63. Terence Brown and Keri Brown, "Ancient DNA and the Archaeologist," *Antiquity* 66 (1992): 10–23; Terence Brown and Keri Brown, "Ancient DNA: Using Molecular Biology to Explore the Past," *BioEssays* 16, no. 10 (1994): 719–26.

64. Linda Vigilant et al., "Mitochondrial DNA Sequences in Single Hairs from a Southern African Population," *Proceedings of the National Academy of Sciences of the United States of America* 86 (1989): 9350–54; Catherine Hänni et al., "Amplification of Mitochondrial DNA Fragments from Ancient Human Teeth and Bones," *Comptes Rendus de l'Academie Des Sciences, Serie III, Sciences de La Vie* 310, no. 9 (1990): 365–70; Susanne Hummel and Bernd Herrmann, "Y-Chromosome-Specific DNA Amplified in Ancient Human Bone," *Naturwissenschaften* 78 (1991): 266–67; Hagelberg and Clegg, "Genetic Polymorphisms in Prehistoric Pacific Islanders"; Anne Stone and Mark

Stoneking, "Ancient DNA from a Pre-Columbian Amerindian Population," *American Journal of Physical Anthropology* 92 (1993): 463–71; Erika Hagelberg et al., "DNA from Ancient Easter Islanders," *Nature* 369 (1994): 25–26; Peter Gill et al., "Identification of the Remains of the Romanov Family by DNA Analysis," *Nature Genetics* 6, no. 2 (1994): 130–35; Oliva Handt et al., "Molecular Genetic Analyses of the Tyrolean Ice Man," *Science* 264, no. 5166 (1994): 1775–78; Marina Faerman et al., "Sex Identification of Archaeological Human Remains Based on Amplification of the X and Y Amelogenin Alleles," *Gene* 167 (1995): 327–32; Marina Faerman et al., "Determining the Sex of Infanticide Victims from the Late Roman Era Through Ancient DNA Analysis," *Journal of Archaeological Science* 25, no. 9 (1998): 861–65.

65. Mark Spigelman and Eshetu Lemma, "The Use of the Polymerase Chain Reaction (PCR) to Detect Mycobacterium Tuberculosis in Ancient Skeletons," *International Journal of Osteoarchaeology* 3, no. 2 (1993): 137–43; A. Rafi et al., "Mycobacterium Leprae DNA from Ancient Bone Detected by PCR," *Lancet* 343, no. 8909 (1994): 1360–61; Wilmar L. Salo et al., "Identification of Mycobacterium Tuberculosis DNA in a Pre-Columbian Peruvian Mummy," *Microbiology* 91 (1994): 2091–94; Heike Baron, Susanne Hummel, and Bernd Herrmann, "Mycobacterium Tuberculosis Complex DNA in Ancient Human Bones," *Journal of Archaeological Science* 23, no. 5 (1996): 667–71.

66. Robert K. Wayne, Jennifer A. Leonard, and Alan Cooper, "Full of Sound and Fury: The Recent History of Ancient DNA," *Annual Review of Ecology and Systematics* 30 (1999): 457–77.

67. Mark Stoneking, "Ancient DNA: How Do You Know When You Have It and What Can You Do with It?" *American Journal of Human Genetics* 57, no. 6 (1995): 1259.

68. Elaine Béraud-Colomb et al., "Human Beta-Globin Gene Polymorphisms Characterized in DNA Extracted from Ancient Bones 12,000 Years Old," *American Journal of Human Genetics* 57, no. 6 (1995): 1267–74.

69. Stoneking, "Ancient DNA: How Do You Know When You Have It and What Can You Do with It?" 1260.

70. Stoneking, "Ancient DNA: How Do You Know When You Have It and What Can You Do with It?" 1260, 1261.

71. Alan Cooper, "Reply to Stoneking: Ancient DNA—How Do You Really Know When You Have It?" *American Journal of Human Genetics* 60 (1997): 1002.

72. Wayne, Leonard, and Cooper, "Full of Sound and Fury," 458–59, 464.

73. Golenberg et al., "Chloroplast DNA Sequence from a Miocene Magnolia Species"; Raúl J. Cano, Hendrik N. Poinar, and George O. Poinar Jr., "Isolation and Partial Characterisation of DNA from the Bee Proplebeia Dominicana (Apidae: Hymenoptera) in 25–40 Million Year Old," *Medical Science Research* 20, no. 7 (1992): 249–51; DeSalle et al., "DNA Sequences from a Fossil Termite"; Cano et al., "Amplification and Sequencing of DNA"; Woodward, Weyand, and Bunnell, "DNA Sequence from Cretaceous Period Bone Fragments."

74. Jackie Fenn and Mark Raskino, *Mastering the Hype Cycle: How to Choose the Right Innovation at the Right Time* (Boston: Harvard Business Press, 2008).

75. Elsbeth Bösl, "Zur Wissenschaftsgeschichte der ADNA-Forschung," *NTM Zeitschrift für Geschichte der Wissenschaften, Technik und Medizin* 25, no. 1 (2017): 99–142; Elsbeth Bösl, *Doing Ancient DNA: Zur Wissenschaftsgeschichte der ADNA-Forschung* (Bielefeld, Germany: Verlag, 2017); Elizabeth Jones and Elsbeth Bösl, "Ancient Human DNA: A History of Hype (Then and Now)," *Journal of Social Archaeology* (February 2021): 1–20.

76. For more information on the role of hype, see work in the sociology of science by Nik Brown, "Hope Against Hype—Accountability in Biopasts, Presents, and Futures," *Science Studies* 16, no. 2 (2003): 3–21; Mads Borup et al., "The Sociology of Expectations in Science and Technology," *Technology Analysis and Strategic Management* 18, nos. 3–4 (2006): 285–98; Harro van Lente, Charlotte Spitters, and Alexander Peine, "Comparing Technological Hype Cycles: Towards a Theory," *Technological Forecasting and Social Change* 80 (2013): 1615–28.

77. Mike Michael, "Futures of the Present: From Performativity to Prehension," in *Contested Futures: A Sociology of Prospective Techno-Science*, ed. Nik Brown, Brian Rapport, and Andrew Webster (Aldershot, U.K.: Ashgate, 2000), 21–42; Brown, "Hope Against Hype"; Borup et al., "The Sociology of Expectations in Science and Technology."

78. Brown, "Hope Against Hype," 11.

79. Brown, "Hope Against Hype," 17, 9.

CHAPTER 6. CONTAMINATION

1. "5th International Ancient DNA Conference," University of Manchester, July 2000; Holger Schutkowski, "5th International Ancient DNA Conference, July 12–14, 2000," *Anthropologischer Anzeiger* 59, no. 2 (2001): 179–81.

2. Erik Stokstad, "Divining Diet and Disease from DNA," *Science* 289, no. 5479 (2000): 530–31.

3. Hendrik N. Poinar et al., "Molecular Coproscopy: Dung and Diet of the Extinct Ground Sloth Nothrotheriops Shastensis," *Science* 281, no. 5375 (1998): 402–6.

4. Alex D. Greenwood et al., "Nuclear DNA Sequences from Late Pleistocene Megafauna," *Molecular Biology and Evolution* 16, no. 11 (1999): 1466–73. Nuclear DNA (nDNA) comes from the nucleus of a cell and contains detailed information about the individual organism. Scientists can use nDNA to identify individuals and determine their genetic relationships to others within and between groups. However, nDNA has a much lower copy number in the cell, which means it is less likely to be preserved and extracted from ancient specimens. Mitochondrial DNA (mtDNA), on the other hand, is found in a cell's mitochondria and has a higher copy number so a higher likelihood that some DNA may be preserved. mtDNA, passed down the maternal line, can be used to differentiate between species as well as determine which individuals are related through their maternal line. See also Michael Hofreiter et al., "Ancient DNA," *Nature Reviews* 2, no. 5 (2001): 353–59.

5. Stokstad, "Divining Diet and Disease from DNA"; Alex D. Greenwood et al., "Evolution of Endogenous Retrovirus-Like Elements of the Woolly Mammoth

(Mammuthus Primigenius) and Its Relatives," *Molecular Biology and Evolution* 18, no. 5 (2001): 840–47.

6. Ross D. E. MacPhee and Preston A. Marx, *The 40,000-Year Plague: Humans, Hyperdisease, and First-Contact Extinctions* (Washington, D.C.: Smithsonian Institution Press, 1997); Stokstad, "Divining Diet and Disease from DNA."

7. Stokstad, "Divining Diet and Disease from DNA," 53.

8. Alan Cooper and Hendrik N. Poinar, "Ancient DNA: Do It Right or Not at All," *Science* 289, no. 5482 (2000): 1139.

9. Svante Pääbo, Russell G. Higuchi, and Allan C. Wilson, "Ancient DNA and the Polymerase Chain Reaction," *Journal of Biological Chemistry* 264, no. 17 (1989): 9709–12; Oliva Handt et al., "Ancient DNA: Methodological Challenges," *Experientia* 50, no. 6 (1994): 524–29.

10. Cooper and Poinar "Ancient DNA: Do It Right or Not at All," 1139.

11. Hofreiter et al., "Ancient DNA"; Svante Pääbo et al., "Genetic Analyses from Ancient DNA," *Annual Review of Genetics* 38, no. 1 (2004): 645–79; Eske Willerslev, Anders J. Hansen, and Hendrik N. Poinar, "Isolation of Nucleic Acids and Cultures from Fossil Ice and Permafrost," *Trends in Ecology and Evolution* 19, no. 3 (2004): 140–47.

12. Svante Pääbo, "Ancient DNA: Extraction, Characterization, Molecular Cloning, and Enzymatic Amplification," *Proceedings of the National Academy of Sciences of the United States of America* 86, no. 6 (1989): 1939–43; Tomas Lindahl, "Instability and Decay of the Primary Structure of DNA," *Nature* 362, no. 6422 (1993): 709–15; Bernd Herrmann and Susanne Hummel, eds., *Ancient DNA: Recovery and Analysis of Genetic Material from Paleontological, Archaeological, Museum, Medical, and Forensic Specimens* (New York: Springer-Verlag, 1994); Handt et al., "Ancient DNA: Methodological Challenges"; Matthias Höss et al., "DNA Damage and DNA Sequence Retrieval from Ancient Tissues," *Nucleic Acids Research* 24, no. 7 (1996): 1304–7; Matthias Krings et al., "Neandertal DNA Sequences and the Origin of Modern Humans," *Cell* 90 (1997): 19–30.

13. Krings et al., "Neandertal DNA Sequences and the Origin of Modern Humans," 19–30.

14. Cooper and Poinar, "Ancient DNA: Do It Right or Not at All," 1139.

15. Pääbo et al., "Genetic Analyses from Ancient DNA."

16. Russell Higuchi et al., "DNA Sequences from the Quagga, an Extinct Member of the Horse Family," *Nature* 312, no. 5991 (1984): 282–84; Svante Pääbo, "Molecular Cloning of Ancient Egyptian Mummy DNA," *Nature* 314, no. 6012 (1985): 644–45.

17. Nick Zagorski, "The Profile of Svante Pääbo," *Proceedings of the National Academy of Sciences of the United States of America* 103, no. 37 (2006): 13575–77.

18. Pääbo et al., "Genetic Analyses from Ancient DNA," 646.

19. For specifics on examples of molecular damage, such as strand breaks, DNA cross-links, as well as oxidative and hydrolytic lesions, see Pääbo et al., "Genetic Analyses from Ancient DNA," 646–54.

20. Pääbo et al., "Genetic Analyses from Ancient DNA," 655.

21. Cooper and Poinar's 2000 publication, "Ancient DNA: Do It Right or Not at All," lists nine key criteria of authenticity, whereas Pääbo and colleagues' 2004 publication, "Genetic Analyses from Ancient DNA," lists eight. Pääbo and colleagues left off the first criteria that ancient DNA studies be conducted in a physically isolated laboratory, but they did discuss the necessity of performing DNA extractions and amplifications in dedicated laboratory facilities where no post-PCR work had been conducted. They also maintained that extraction work should be conducted with protective clothing and the workspace cleaned regularly with oxidants such as bleach and irradiated with UV lights. Also, instead of requiring that ancient DNA sequences be recovered from associated remains of the specimen under investigation, they stated that studies must test for the occurrence of nuclear insertions of mtDNA.

22. Pääbo et al., "Genetic Analyses from Ancient DNA," 659.

23. Hofreiter et al., "Ancient DNA."

24. Pääbo et al., "Genetic Analyses from Ancient DNA," 670.

25. Harry M. Collins and Trevor J. Pinch, "The Construction of the Paranormal: Nothing Unscientific Is Happening," in *On the Margins of Science: The Social Construction of Rejected Knowledge*, ed. Roy Wallis, *Sociological Review Monograph* 27 (1979): 237–70; Harry Collins, *Changing Order: Replication and Induction in Scientific Practice* (Chicago: University of Chicago Press, 1985); Harry Collins and Trevor Pinch, *The Golem: What You Should Know About Science* (Cambridge: Cambridge University Press, 1993); Harry Collins, *Gravity's Shadow: The Search for Gravitational Waves* (Chicago: University of Chicago Press, 2004); Harry Collins, *Gravity's Ghost: Scientific Discovery in the Twenty-First Century* (Chicago: University of Chicago Press, 2010).

26. Collins, *Changing Order*, 28.

27. Harry M. Collins, "Son of Seven Sexes: The Social Destruction of a Physical Phenomenon," *Social Studies of Science* 11, no. 1 (1981): 34.

28. Collins, *Changing Order*, 2.

29. Harry M. Collins, "Reproducibility of Experiments: Experimenters' Regress, Statistical Uncertainty Principle, and the Replication Imperative," in *Reproducibility: Principles, Problems, Practices, and Prospects*, ed. Harald Atmanspacher and Sabine Maasen (Hoboken, N.J.: Wiley, 2016), 66.

30. Collins, *Changing Order*, 2.

31. Collins, "Reproducibility of Experiments," 66.

32. Thomas F. Gieryn, "Boundary-Work and the Demarcation of Science from Non-Science: Strains and Interests in Professional Ideologies of Scientists," *American Sociological Review* 48, no. 6 (1983): 781–95.

33. Thomas F. Gieryn, *Cultural Boundaries of Science: Credibility on the Line* (Chicago: University of Chicago Press, 1999), 4–5.

34. Gieryn, *Cultural Boundaries of Science*, 16.

35. Gieryn, *Cultural Boundaries of Science*, 63.

36. Rob DeSalle and David Lindley, *The Science of* Jurassic Park *and* The Lost World; *or, How to Build a Dinosaur* (New York: Basic, 1997); David A. Kirby, *Lab Coats in Hollywood: Science, Scientists, and Cinema* (Cambridge, Mass.: MIT Press, 2013).

37. David Norman, "Misread in Tooth and Claw," *Times Higher Education,* November 28, 1997, 22.

38. Adrian M. Lister, "Ancient DNA: Not Quite *Jurassic Park,*" *Trends in Ecology and Evolution* 9, no. 3 (1994): 82–84; Mary Schweitzer and Tracy Staedter, "The Real Jurassic Park," *Earth* (June 1997): 55–57.

39. In describing this division, it is important to note that the line between these factions was not necessarily hard and fast. Instead, it was permeable, especially as some scientists have tried to collaborate across the schism. Indeed, today some do collaborate despite past controversy. Regardless, descriptions of "believers" versus "non-believers" in their oral-history testimonies has helped scientists make sense of an important issue, concerns about contamination, and its effect on their field. Adhering to this division and their commitment to the retelling of it (to colleagues, collaborators, students, and historians like me) serves to reinforce the seriousness of these contamination concerns and the consequences they felt it had on their credibility. To be clear, this map of divisions is by no means the only map of interactions that interviewees tried to draw regarding the history of this community. However, it is one way that the majority of interviewees have described this time of their history, and this distinction has served to set up the field's growth in the years to come.

40. Erika Hagelberg referred to these terms in her talk on the history of the discipline, "Ancient DNA: The First Three Decades," paper presented at the Royal Society, London, November 2013. Charles Greenblatt and Bernd Herrmann also used these terms to describe the divide in an unpublished chapter for an unpublished book. See Bernd Herrmann and Charles Greenblatt, "A Short Essay on ADNA and Its Future," 2010, Author's Personal Collection (file from Bernd Herrmann and Charles Greenblatt).

41. "The 6th International Conference on Ancient DNA and Associated Biomolecules," Tel Aviv, Israel, July 21–25, 2002.

42. Mark Spigelman and Eshetu Lemma, "The Use of the Polymerase Chain Reaction (PCR) to Detect Mycobacterium Tuberculosis in Ancient Skeletons," *International Journal of Osteoarchaeology* 3, no. 2 (1993): 137–43; A. Rafi et al., "Mycobacterium Leprae DNA from Ancient Bone Detected by PCR," *Lancet* 343, no. 8909 (1994): 1360–61; Wilmar L. Salo et al., "Identification of Mycobacterium Tuberculosis DNA in a Pre-Columbian Peruvian Mummy," *Microbiology* 91 (1994): 2091–94; Heike Baron, Susanne Hummel, and Bernd Herrmann, "Mycobacterium Tuberculosis Complex DNA in Ancient Human Bones," *Journal of Archaeological Science* 23, no. 5 (1996): 667–71.

43. Robert K. Wayne, Jennifer A. Leonard, and Alan Cooper, "Full of Sound and Fury: The Recent History of Ancient DNA," *Annual Review of Ecology and Systematics* 30 (1999): 467–68.

44. "The First International Symposium on Biomolecular Archaeology," Amsterdam, Netherlands, March 18–20, 2018.

45. Herrmann and Greenblatt, "A Short Essay on ADNA and Its Future," 2.

46. Herrmann and Greenblatt, "A Short Essay on ADNA and Its Future," 2.

47. Herrmann and Hummel, eds., *Ancient DNA*; Susanne Hummel, ed., *Ancient DNA Typing: Methods, Strategies, and Applications* (Berlin: Springer, 2003); Helen D. Donoghue et al., "Tuberculosis: From Prehistory to Robert Koch, as Revealed by Ancient DNA," *Lancet: Infectious Diseases* 4, no. 9 (September 2004): 584–92; Pääbo et al., "Genetic Analyses from Ancient DNA"; Eske Willerslev and Alan Cooper, "Ancient DNA," *Proceedings of the Royal Society, Series B, Biological Sciences* 272, no. 1558 (2005): 3–16; Alicia K. Wilbur et al., "Deficiencies and Challenges in the Study of Ancient Tuberculosis DNA," *Journal of Archaeological Science* 36, no. 9 (September 2009): 1990–97; Beth Shapiro and Michael Hofreiter, eds., *Ancient DNA: Methods and Protocols* (New York: Springer, 2012).
48. Willerslev and Cooper, "Ancient DNA," 3.
49. M. Thomas P. Gilbert et al., "Assessing Ancient DNA Studies," *Trends in Ecology and Evolution* 20, no. 10 (2005): 541.
50. Gilbert et al., "Assessing Ancient DNA Studies," 542.
51. Gilbert et al., "Assessing Ancient DNA Studies," 542.
52. Pääbo et al., "Genetic Analyses from Ancient DNA"; Willerslev and Cooper, "Ancient DNA."
53. Collins and Pinch, *The Golem*, 3.
54. Collins, *Changing Order*, 19.
55. Gieryn, *Cultural Boundaries of Science*, 63.
56. Peter Broks, *Understanding Popular Science* (Maidenhead, U.K.: Open University Press, 2006), 107.

CHAPTER 7. ANCIENT GENETICS TO ANCIENT GENOMICS

1. Marcel Margulies et al., "Genome Sequencing in Microfabricated High-Density Picolitre Reactors," *Nature* 437, no. 7057 (2005): 376–80.
2. Jonathan M. Rothberg and John H. Leamon, "The Development and Impact of 454 Sequencing," *Nature Biotechnology* 26, no. 10 (2008): 1123.
3. Karl V. Voelkerding, Shale A. Dames, and Jacob D. Durtschi, "Next-Generation Sequencing: From Basic Research to Diagnostics," *Clinical Chemistry* 55, no. 4 (2009): 461–62.
4. Roche (454) GS FLX is a technology based on a method known as parallel pyro-sequencing, while the Illumina (Solexa) Genome Analyzer uses a method based on reversible terminators. The 454 technology generates longer reads of DNA (over 400 base pairs) but is somewhat error-prone in homopolymeric regions (e.g., CCCCCC). Illumina generates shorter reads of DNA (100–150 base pairs) but in greater numbers. Both technologies operate on the same rationale for the production of sequences. However, they differ in the amplification procedures and sequencing chemistry, resulting in different throughputs. See Ermanno Rizzi et al., "Ancient DNA Studies: New Perspectives on Old Samples," *Genetics, Selection, Evolution* 44, no. 1 (2012): 1–19.
5. James P. Noonan et al., "Genomic Sequencing of Pleistocene Cave Bears," *Science* 309, no. 5734 (2005): 597–600.
6. Hendrik N. Poinar et al., "Metagenomics to Paleogenomics: Large-Scale Sequencing of Mammoth DNA," *Science* 311, no. 2006 (2006): 393.

7. Michael Knapp and Michael Hofreiter, "Next Generation Sequencing of Ancient DNA: Requirements, Strategies and Perspectives," *Genes* 1, no. 2 (2010): 227–43.

8. Alan Cooper, "The Year of the Mammoth," *PLoS Biology* 4, no. 3 (2006): 0311–0313.

9. Evgeny I. Rogaev et al., "Complete Mitochondrial Genome and Phylogeny of Pleistocene Mammoth Mammuthus Primigenius," *PLoS Biology* 4, no. 3 (2006): 0403–10; Johannes Krause et al., "Multiplex Amplification of the Mammoth Mitochondrial Genome and the Evolution of Elephantidae," *Nature* 439, no. 7077 (2006): 724–27; Mark G. Thomas, Neil Bradman, and Helen M. Flinn, "High Throughput Analysis of 10 Microsatellite and 11 Diallelic Polymorphisms on the Human Y-Chromosome," *Human Genetics* 105, no. 6 (1999): 577–81.

10. Poinar et al., "Metagenomics to Paleogenomics."

11. Cooper, "The Year of the Mammoth," 0313.

12. Didier Raoult et al., "Molecular Identification by 'Suicide PCR' of Yersinia Pestis as the Agent of Medieval Black Death," *Proceedings of the National Academy of Sciences of the United States of America* 97, no. 23 (2000): 12800–803.

13. M. Thomas P. Gilbert et al., "Absence of Yersinia Pestis-Specific DNA in Human Teeth from Five European Excavations of Putative Plague Victims," *Microbiology* 150 (2004): 341–54.

14. Kirsten Bos et al., "A Draft Genome of Yersinia Pestis from Victims of the Black Death," *Nature* 478, no. 7370 (2011): 506–10.

15. Alan Cooper et al., "Complete Mitochondrial Genome Sequences of Two Extinct Moas Clarify Ratite Evolution," *Nature* 409, no. 6821 (2001): 704–7; Oliver Haddrath and Allan J. Baker, "Complete Mitochondrial DNA Genome Sequences of Extinct Birds: Ratite Phylogenetics and the Vicariance Biogeography Hypothesis," *Proceedings of the Royal Society, Series B, Biological Sciences* 268, no. 1470 (2001): 939–45; Ann H. Reid et al., "Characterization of the 1918 'Spanish' Influenza Virus Matrix Gene Segment," *Journal of Virology* 76, no. 21 (2002): 10717–23; Terrence M. Tumpey et al., "Characterization of the Reconstructed 1918 Spanish Influenza Pandemic Virus," *Science* 310, no. 5745 (2005): 77–80; Noonan et al., "Genomic Sequencing of Pleistocene Cave Bears"; M. Thomas P. Gilbert et al., "Characterization of Genetic Miscoding Lesions Caused by Postmortem Damage," *American Journal of Human Genetics* 72, no. 1 (2003): 48–61; M. Thomas P. Gilbert et al., "Distribution Patterns of Postmortem Damage in Human Mitochondrial DNA," *American Journal of Human Genetics* 72, no. 1 (2003): 32–47.

16. Jennifer A. Leonard, Robert K. Wayne, and Alan Cooper, "Population Genetics of Ice Age Brown Bears," *Proceedings of the National Academy of Sciences of the United States of America* 97, no. 4 (2000): 1651–54.

17. Beth Shapiro et al., "Rise and Fall of the Beringian Steppe Bison," *Science* 306, no. 5701 (2004): 1561–65.

18. Odile Loreille et al., "Ancient DNA Analysis Reveals Divergence of the Cave Bear, Ursus Spelaeus, and Brown Bear, Ursus Arctos, Lineages," *Current Biology* 11, no. 3 (2001): 200–203; Carles Vilà et al., "Widespread Origins of Domestic Horse Lineages," *Science* 291, no. 5503 (2001): 474–77; Ian Barnes et al., "Dynamics of Pleistocene Population Extinctions in Beringian Brown Bears,"

Science 295, no. 5563 (2002): 2267–70; Michael Hofreiter et al., "Ancient DNA Analyses Reveal High Mitochondrial DNA Sequence Diversity and Parallel Morphological Evolution of Late Pleistocene Cave Bears," *Molecular Biology and Evolution* 19, no. 8 (2002): 1244–50; Jennifer A. Leonard et al., "Ancient DNA Evidence for Old World Origin of New World Dogs," *Science* 298, no. 5598 (2002): 1613–16; Peter A. Ritchie et al., "Ancient DNA Enables Timing of the Pleistocene Origin and Holocene Expansion of Two Adelie Penguin Lineages in Antarctica," *Molecular Biology and Evolution* 21, no. 2 (2003): 240–48; Michael Hofreiter et al., "Evidence for Reproductive Isolation Between Cave Bear Populations," *Current Biology* 14, no. 1 (2004): 40–43; Shapiro et al., "Rise and Fall of the Beringian Steppe Bison."

19. "Neandertal Genome to Be Deciphered," Max Planck Society, July 20, 2006, www.mpg.de/534422/pressRelease20060720.

20. Svante Pääbo, *Neanderthal Man: In Search of Lost Genomes* (New York: Basic, 2014), 124.

21. Pääbo, *Neanderthal Man*, 117, 124.

22. Jeremy Schmutz et al., "Quality Assessment of the Human Genome Sequence," *Nature* 429, no. 6990 (2004): 365–68; "Timeline: Organisms That Have Had Their Genomes Sequenced," Your Genome, www.yourgenome.org/facts/timeline-organisms-that-have-had-their-genomes-sequenced.

23. Dorothy Nelkin and Susan Lindee, *The DNA Mystique: The Gene as a Cultural Icon* (New York: Freeman, 1995); Suzanne Anker and Dorothy Nelkin, *The Molecular Gaze: Art in the Genetic Age* (Cold Spring Harbor, N.Y.: Cold Spring Harbor Laboratory Press, 2003); Dorothy Nelkin, "Molecular Metaphors: The Gene in Popular Discourse," *Nature Reviews Genetics* 2, no. 7 (2001): 555–59; Elsbeth Bösl, "Zur Wissenschaftsgeschichte der ADNA-Forschung," *NTM Zeitschrift für Geschichte der Wissenschaften, Technik und Medizin* 25, no. 1 (2017): 99–142; Elsbeth Bösl, *Doing Ancient DNA: Zur Wissenschaftsgeschichte der ADNA-Forschung* (Bielefeld, Germany: Verlag, 2017).

24. Matthias Krings et al., "Neandertal DNA Sequences and the Origin of Modern Humans," *Cell* 90 (1997): 19–30; Igor Ovchinnikov et al., "Molecular Analysis of Neanderthal DNA from the Northern Caucasus," *Nature* 404, no. 6777 (2000): 490–93; Matthias Höss, "Neanderthal Population Genetics," *Nature* 404, no. 6777 (2000): 453–54.

25. Krings et al., "Neandertal DNA Sequences and the Origin of Modern Humans."

26. Richard E. Green et al., "Analysis of One Million Base Pairs of Neanderthal DNA," *Nature* 444, no. 7117 (2006): 330–36; James P. Noonan et al., "Sequencing and Analysis of Neanderthal Genomic DNA," *Science* 314, no. 5802 (2006): 1113–18; Pääbo, *Neanderthal Man*, chap. 11.

27. Pääbo, *Neanderthal Man*, chap. 11.

28. Jeffrey D. Wall and Sung K. Kim, "Inconsistencies in Neanderthal Genomic DNA Sequences," *PLoS Genetics* 3, no. 10 (2007): 1865.

29. Pääbo, *Neanderthal Man*, chap. 11.

30. Rex Dalton, "DNA Probe Finds Hints of Human," *Nature* 449 (September 6, 2007): 7.

31. Pääbo, *Neanderthal Man,* chap. 11.

32. Annalee Newitz, "Code of the Caveman," *Wired,* July 2006, www.wired.com /2006/07/caveman/.

33. Pääbo, *Neanderthal Man,* chap. 17.

34. Richard E. Green et al., "A Draft Sequence of the Neandertal Genome," *Science* 328, no. 5979 (2010): 710–22.

35. "The Neandertal in Us," Max Planck Society, 2010, www.mpg.de/617258/press-Release20100430; Ewen Callaway, "Neanderthal Genome Reveals Interbreeding with Humans," *New Scientist,* May 6, 2010, www.newscientist.com/article/ dn18869-neanderthal-genome-reveals-interbreeding-with-humans/; Pääbo, *Neanderthal Man.*

36. Ker Than, "Neanderthals, Humans Interbred—First Solid DNA Evidence," *National Geographic,* May 8, 2010, http://news.nationalgeographic.com/news /2010/05/100506-science-neanderthals-humans-mated-interbred-dna-gene/.

37. "Neanderthal," Answers in Genesis, https://answersingenesis.org/human-evolution/neanderthal/.

38. Pääbo, *Neanderthal Man,* 221–22.

39. M. Thomas P. Gilbert et al., "Paleo-Eskimo MtDNA Genome Reveals Matrilineal Discontinuity in Greenland," *Science* 320, no. 5884 (2008): 1787–89; M. Thomas P. Gilbert et al., "Intraspecific Phylogenetic Analysis of Siberian Woolly Mammoths Using Complete Mitochondrial Genomes," *Proceedings of the National Academy of Sciences of the United States of America* 105, no. 24 (2008): 8327–32; Webb Miller et al., "Sequencing the Nuclear Genome of the Extinct Woolly Mammoth," *Nature* 456, no. 7220 (2008): 387–90; Morten Rasmussen et al., "Ancient Human Genome Sequence of an Extinct Palaeo-Eskimo," *Nature* 463, no. 7282 (2010): 757–62; Morten Rasmussen et al., "An Aboriginal Australian Genome Reveals Separate Human Dispersals into Asia," *Science* 334, no. 6052 (2011): 94–98; Jakob Skou Pedersen et al., "Genome-Wide Nucleosome Map and Cytosine Methylation Levels of an Ancient Human Genome," *Genome Research* 24, no. 3 (2014): 454–66; Morten Rasmussen et al., "The Genome of a Late Pleistocene Human from a Clovis Burial Site in Western Montana," *Nature* 506, no. 7487 (2014): 225–29; Eske Willerslev et al., "Fifty Thousand Years of Arctic Vegetation and Megafaunal Diet," *Nature* 506, no. 7486 (2014): 47–51; Turi E. King et al., "Identification of the Remains of King Richard III," *Nature Communications* 5 (2014): 5631.

40. N. Izagirre and C. de la Rúa, "An MtDNA Analysis in Ancient Basque Populations: Implications for Haplogroup V as a Marker for a Major Paleolithic Expansion from Southwestern Europe," *American Journal of Human Genetics* 65, no. 1 (1999): 199–207; Wolfgang Haak et al., "Ancient DNA from the First European Farmers in 7500-Year-Old Neolithic Sites," *Science* 310, no. 5750 (2005): 1016–18; Joachim Burger et al., "Absence of the Lactase-Persistence-Associated Allele in Early Neolithic Europeans," *Proceedings of the National Academy of Sciences of the United States of America* 104, no. 10 (2007): 3736–41; B. Bramanti et al., "Genetic Discontinuity Between Local Hunter-Gatherers and Central Europe's First Farmers," *Science* 326, no. 5949 (2009): 137–40;

Wolfgang Haak et al., "Ancient DNA from European Early Neolithic Farmers Reveals Their Near Eastern Affinities," *PLoS Biology* 8, no. 11 (2010): 1–16; Pontus Skoglund et al., "Origins and Genetic Legacy of Neolithic Farmers and Hunter-Gatherers in Europe," *Science* 336, no. 6080 (April 27, 2012): 466–69; Christina Warinner et al., "Pathogens and Host Immunity in the Ancient Human Oral Cavity," *Nature Genetics* 46, no. 4 (2014): 336–44; Eppie R. Jones et al., "Upper Palaeolithic Genomes Reveal Deep Roots of Modern Eurasians," *Nature Communications* 6 (November 16, 2015): 8912; Helena Malmström et al., "Ancient Mitochondrial DNA from the Northern Fringe of the Neolithic Farming Expansion in Europe Sheds Light on the Dispersion Process," *Philosophical Transactions of the Royal Society of London, Series B, Biological Sciences* 370, no. 1660 (January 19, 2015): 1–10.

41. Leonard et al., "Ancient DNA Evidence for Old World Origin of New World Dogs"; Ruth Bollongino et al., "Early History of European Domestic Cattle as Revealed by Ancient DNA," *Biology Letters* 2, no. 1 (2006): 155–159; G. Larson et al., "Ancient DNA, Pig Domestication, and the Spread of the Neolithic into Europe," *Proceedings of the National Academy of Sciences of the United States of America* 104, no. 39 (September 25, 2007): 15276–81; Amelie Scheu et al., "Ancient DNA Provides No Evidence for Independent Domestication of Cattle in Mesolithic Rosenhof, Northern Germany," *Journal of Archaeological Science* 35, no. 5 (2008): 1257–1264; Greger Larson et al., "Rethinking Dog Domestication by Integrating Genetics, Archeology, and Biogeography," *Proceedings of the National Academy of Sciences of the United States of America* 109, no. 23 (June 5, 2012): 8878–83; Greger Larson et al., "Current Perspectives and the Future of Domestication Studies," *Proceedings of the National Academy of Sciences of the United States of America* 111, no. 17 (2014): 6139–46; Pontus Skoglund et al., "Ancient Wolf Genome Reveals an Early Divergence of Domestic Dog Ancestors and Admixture into High-Latitude Breeds," *Current Biology* 25 (2015): 1–5.

42. Green et al., "A Draft Sequence of the Neandertal Genome."

43. Johannes Krause et al., "A Complete MtDNA Genome of an Early Modern Human from Kostenki, Russia," *Current Biology*, vol. 20 (2010): 231–36; Johannes Krause et al., "The Complete Mitochondrial DNA Genome of an Unknown Hominin from Southern Siberia," *Nature* 464, no. 7290 (2010): 894–97; David Reich et al., "Genetic History of an Archaic Hominin Group from Denisova Cave in Siberia," *Nature* 468, no. 7327 (2010): 1053–60; Ann Gibbons, "A Crystal-Clear View of an Extinct Girl's Genome," *Science* 337 (2012): 1028–29; David Gokhman et al., "Reconstructing the DNA Methylation Maps of the Neandertal and the Denisovan," *Science* 344, no. 6183 (2014): 523–27.

44. Mark Stoneking and Johannes Krause, "Learning About Human Population History from Ancient and Modern Genomes," *Nature Reviews Genetics* 12, no. 9 (2011): 603–14; Chris Stringer, *Lone Survivors: How We Came to Be the Only Humans on Earth* (New York: St. Martin's, 2012); Krishna R. Veeramah and Michael F. Hammer, "The Impact of Whole-Genome Sequencing on the Reconstruction of Human Population History," *Nature Reviews* 15 (2014):

149–62; Ann Gibbons, "Revolution in Human Evolution," *Science* 349, no. 6246 (2015): 362–66.

45. Margulies et al., "Genome Sequencing in Microfabricated High-Density Picolitre Reactors"; Craig D. Millar et al., "New Developments in Ancient Genomics," *Trends in Ecology and Evolution* 23, no. 7 (2008): 386–93; Knapp and Hofreiter, "Next Generation Sequencing of Ancient DNA."

46. Nick Zagorski, "The Profile of Svante Pääbo," *Proceedings of the National Academy of Sciences of the United States of America* 103, no. 37 (2006): 13575–77; Carl Zimmer, "Eske Willerslev Is Rewriting History with DNA," *New York Times,* May 17, 2016, www.nytimes.com/2016/05/17/science/eske-willerslev-ancient-dna-scientist.html.

47. Gideon Lewis, "Is Ancient DNA Research Revealing New Truths—or Falling Into Old Traps?" *New York Times,* January 17, 2019, www.nytimes.com /2019/01/17/magazine/ancient-dna-paleogenomics.html.

48. David Reich, *Who We Are and How We Got Here: Ancient DNA and the New Science of the Human Past* (New York: Pantheon, 2018), xxiv.

49. David Reich, "How to Talk About 'Race' and Genetics," *New York Times,* March 30, 2018, www.nytimes.com/2018/03/30/opinion/race-genetics.html; Jonathan Kahn et al., "How Not To Talk About Race and Genetics," Center for Genetics and Society, 2018, www.geneticsandsociety.org/article/how-not -talk-about-race-and-genetics.

50. Alexandra Ion, "How Interdisciplinary Is Interdisciplinary? Revisiting the Impact of ADNA Research for the Archaeology of Human Remains," *Current Swedish Archaeology* 25 (2017): 87–108.

51. See Elizabeth D. Jones and Elsbeth Bösl, "Ancient Human DNA: A History of Hype (Then and Now)," *Journal of Social Archaeology* (February 2021): 1–20. See also Thomas Booth, "A Stranger in a Strange Land: A Perspective on Archaeological Responses to the Palaeogenetic Revolution from an Archaeologist Working Amongst Palaeogeneticists," *World Archaeology* 51 (2019): 586–601; Craig D. Millar and D. Michael Lambert, "Archaeogenetics and Human Evolution: The Ontogeny of a Biological Discipline," *World Archaeology* 51 (2019): 546–59; Catherine Frieman and Daniela Hofmann, "Present Pasts in the Archaeology of Genetics, Identity, and Migration in Europe: A Critical Essay," *World Archaeology* 51 (2019): 528–45; Susanne Hakenbeck, "Genetics, Archaeology and the Far Right: An Unholy Trinity," *World Archaeology* 51 (2019): 517–27; Michael L. Blakey, "On the Biodeterministic Imagination," *Archaeological Dialogues* 27 (2020): 1–16; Martin Furholt, "Biodeterminism and Pseudo-Objectivity as Obstacles for the Emerging Field of Archaeogenetics," *Archaeological Dialogues* 27 (2020): 23–25; Thomas Booth, "Imagined Biodeterminism?" *Archaeological Dialogues* 27 (2020): 16–19; Ion, "How Interdisciplinary Is Interdisciplinary?"

52. Rachel J. Crellin and Oliver J. T. Harris, "Beyond Binaries: Interrogating Ancient DNA," *Archaeological Dialogues* 27 (2020): 37–56.

53. Ewen Callaway, "The Battle for Common Ground," *Nature* 555, no. 7698 (2018): 574.

54. Blakey, "On the Biodeterministic Imagination."

55. Reich, *Who We Are and How We Got Here,* chap. 10.

56. Furholt, "Biodeterminism and Pseudo-Objectivity as Obstacles for the Emerging Field of Archaeogenetics."

57. Blakey, "On the Biodeterministic Imagination."

58. Jenny Reardon, *Race to the Finish: Identity and Governance in an Age of Genomics* (Princeton, N.J.: Princeton University Press, 2004); Jenny Reardon, "Decoding Race and Human Difference in a Genomic Age," *Differences: A Journal of Feminist Cultural Studies* 15, no. 3 (2004): 38–65; Jenny Reardon and Kim TallBear, "'Your DNA Is Our History': Genomics, Anthropology, and the Contruction of Whitness as Property," *Current Anthropology* 53, no. 5 (2012): S233–45; Kim TallBear, "Genomic Articulations of Indigeneity," *Social Studies of Science* 43, no. 4 (2013): 509–33; Kim TallBear, *Native American DNA: Tribal Belonging and the False Promise of Genetic Science* (Minneapolis: University of Minnesota Press, 2013); Joanna Radin, *Life on Ice: A History of New Uses for Cold Blood* (Chicago: University of Chicago Press, 2017).

59. Reardon, *Race to the Finish,* 7.

60. Frieman and Hofmann, "Present Pasts in the Archaeology of Genetics, Identity, and Migration in Europe"; Hakenbeck, "Genetics, Archaeology and the Far Right."

61. Reich, *Who We Are and How We Got Here,* xvii.

62. Maria C. Ávila Arcos, "Troubling Traces of Biocolonialism Undermine an Otherwise Eloquent Synthesis of Ancient Genome Research," *Science,* April 17, 2018, http://blogs.sciencemag.org/books/2018/04/17/who-we-are-and-how -we-got-here/.

63. Ávila Arcos, "Troubling Traces of Biocolonialism"; Reich, *Who We Are and How We Got Here.*

64. Lewis, "Is Ancient DNA Research Revealing New Truths—or Falling Into Old Traps?"

65. Clio Der Sarkissian et al., "Ancient Genomics," *Philosophical Transactions of the Royal Society of London, Series B, Biological Sciences* 370, no. 1660 (January 19, 2015): 1–12, https://doi.org/10.1098/rstb.2013.0387.

66. Anna Källén et al., "Archaeogenetics in Popular Media: Contemporary Implications of Ancient DNA," *Current Swedish Archaeology* 27 (2019): 69–91; C. Hedenstierna-Jonson et al., "A Female Viking Warrior Confirmed by Genomics," *American Journal of Physical Anthropology* 164 (2018): 853–60.

67. Frieman and Hofmann, "Present Pasts in the Archaeology of Genetics, Identity, and Migration in Europe."

CHAPTER 8. CELEBRITY AS IDENTITY

1. Erika Hagelberg, Michael Hofreiter, and Christine Keyser, "Introduction—Ancient DNA: The First Three Decades," *Philosophical Transactions of the Royal Society of London, Series B, Biological Sciences* 370, no. 1660 (2015): 1–6.

2. Vassiliki Betty Smocovitis, "The 1959 Darwin Centennial Celebration in America," *Osiris* 14 (1999): 274–323.

3. Hagelberg, Hofreiter, and Keyser, "Introduction—Ancient DNA."

4. Michael Knapp and Michael Hofreiter, "Next Generation Sequencing of Ancient DNA: Requirements, Strategies and Perspectives," *Genes* 1, no. 2 (2010): 227.

5. Henry Nicholls, "Ancient DNA Comes of Age," *PLoS Biology* 3, no. 2 (February 15, 2005): e56, https://doi.org/10.1371/journal.pbio.0030056.

6. Joseph Allen Cain, "Common Problems and Cooperative Solutions: Organizational Activities in Evolutionary Studies, 1937–1946," *Isis* 84, no. 1 (1993): 1–25.

7. Elsbeth Bösl, "Zur Wissenschaftsgeschichte der ADNA-Forschung," *NTM Zeitschrift für Geschichte der Wissenschaften, Technik und Medizin* 25, no. 1 (2017): 99–142; Elsbeth Bösl, *Doing Ancient DNA: Zur Wissenschaftsgeschichte der ADNA-Forschung* (Bielefeld, Germany: Verlag, 2017).

8. Elisabeth S. Clemmens, "Of Asteroids and Dinosaurs: The Role of the Press in the Shaping of Scientific Debate," *Social Studies of Science* 16 (1986): 421–56; Elisabeth S. Clemmens, "The Impact Hypothesis and Popular Science: Conditions and Consequences of Interdisciplinary Debate," in *The Mass-Extinction Debates: How Science Works in a Crisis,* ed. William Glen (Stanford, Calif.: Stanford University Press, 1994), 92–120.

9. Clemmens, "The Impact Hypothesis and Popular Science," 111, 119.

10. Michael Strevens, "The Role of the Priority Rule in Science," *Philosophy of Science* 100, no. 2 (2003): 55–79.

11. Joe Cain, "Ritual Patricide: Why Stephen Jay Gould Assassinated George Gaylord Simpson," in *The Paleobiological Revolution: Essays on the Growth of Modern Paleontology,* ed. David Sepkoski and Michael Ruse (Chicago: University of Chicago Press, 2009), 252–53.

12. Hagelberg, Hofreiter, and Keyser, "Introduction—Ancient DNA."

13. Bösl, *Doing Ancient DNA.*

14. Ann Gibbons, "Ancient DNA Divide," *Science* 352, no. 6292 (2016): 1384.

15. Bösl, "Zur Wissenschaftsgeschichte der ADNA-Forschung."

16. Bernd Herrmann and Charles Greenblatt, "A Short Essay on ADNA and Its Future," 2010, 3–4, Author's Personal Collection (file from Bernd Herrmann and Charles Greenblatt).

17. Peter Galison, "Computer Simulations and the Trading Zone," in *The Disunity of Science: Boundaries, Contexts, and Power,* ed. Peter Galison and David J. Stump (Stanford, Calif.: Stanford University Press, 1996), 118–57.

18. Elsbeth Bösl, "Zur Wissenschaftsgeschichte der ADNA-Forschung," *NTM Zeitschrift für Geschichte der Wissenschaften, Technik und Medizin* 25, no. 1 (2017): 99–142.

19. Matthew Collins, "Archaeology and the Biomolecular 'Revolution': Too Much of the Wrong Kind of Data," *Stichting Voor de Nederlandse Archeologie* 18 (2006): 1–18.

CHAPTER 9. CELEBRITY AS STRATEGY

1. Erika Hagelberg, Michael Hofreiter, and Christine Keyser, "Introduction—Ancient DNA: The First Three Decades," *Philosophical Transactions of the Royal Society of London, Series B, Biological Sciences* 370, no. 1660 (2015): 1–6.

2. Russell Higuchi et al., "DNA Sequences from the Quagga, an Extinct Member of the Horse Family," *Nature* 312, no. 5991 (1984): 282–84.

3. Rob DeSalle et al., "DNA Sequences from a Fossil Termite in Oligo-Miocene Amber and Their Phylogenetic Implications," *Science* 257, no. 5078 (1992): 1933–36.

4. Morten E. Allentoft et al., "Population Genomics of Bronze Age Eurasia," *Nature* 522, no. 7555 (2015): 167–72.

5. Ewen Callaway, "DNA Data Explosion Lights Up the Bronze Age," *Nature* 522, no. 7555 (June 11, 2015): 140–41.

6. Iñigo Olalde et al., "The Beaker Phenomenon and the Genomic Transformation of Northwest Europe," *Nature* 555, no. 7695 (2018): 190–96; Iain Mathieson et al., "The Genomic History of Southeastern Europe," *Nature* 555, no. 7695 (2018): 197–203.

7. This shift from "too little data" to "too much data" is meant to be understood as a comparison between ancient DNA research's past and present. Even if researchers are able to produce more data, comparatively speaking, the data is still often of poor or patchy quality. This requires researchers to find ways of handling and analyzing the data. Today, there is much more data than before, but the amount of ancient DNA data, if compared to the influx of modern DNA data, is still far off.

8. Adrian M. Lister, "Ancient DNA: Not Quite *Jurassic Park*," *Trends in Ecology and Evolution* 9, no. 3 (1994): 82–84; Rob DeSalle and David Lindley, *The Science of* Jurassic Park *and* The Lost World; *or, How to Build a Dinosaur* (New York: Basic, 1997); Alan Cooper and Hendrik N. Poinar, "Ancient DNA: Do It Right or Not at All," *Science* 289, no. 5482 (2000): 1139; Svante Pääbo et al., "Genetic Analyses from Ancient DNA," *Annual Review of Genetics* 38, no. 1 (2004): 645–79; Martin B. Hebsgaard, Matthew J. Phillips, and Eske Willerslev, "Geologically Ancient DNA: Fact or Artefact?" *Trends in Microbiology* 13, no. 5 (2005): 212–20.

9. I cannot fully do justice to the scientific and philosophical elements that the demarcation discussion requires. However, I do intend to raise some points of connection between this history of ancient DNA research and what philosophers of science have said about the nature and implications of contemporary data-driven inquiry. The history of ancient DNA research has demonstrated the degree to which scientists in search of DNA from fossils were very much driven by data in terms of access to samples and availability of technology. It has also demonstrated that scientists, as well as major journals such as *Nature* and *Science*, were motivated by the celebrity that surrounded the science of ancient DNA research.

10. Karl Popper, *The Logic of Scientific Discovery* (London: Routledge, 1959); Thomas Kuhn, *The Structure of Scientific Revolutions* (Chicago: University of Chicago Press, 1962); Imre Lakatos, "Falsification and the Methodology of Research Program," in *Criticism and the Growth of Knowledge*, ed. Imre Lakatos and Alan Musgrave (Cambridge: Cambridge University Press, 1970), 91–197.

11. Maureen A. O'Malley et al., "Philosophies of Funding," *Cell* 138, no. 4 (2009): 611–15; Chris Haufe, "Why Do Funding Agencies Favor Hypothesis Testing?" *Studies in History and Philosophy of Science* 44 (2013): 363–74.

12. Lakatos, "Falsification and the Methodology of Research Program"; Paul Feyerabend, *Against Method* (London: New Left, 1975); Larry Laudan, *Science and Hypothesis: Historical Essays on Scientific Methodology* (London: Reidel, 1981).

13. Sabina Leonelli, "Introduction: Making Sense of Data-Driven Research in the Biological and Biomedical Sciences," *Studies in History and Philosophy of Biological and Biomedical Sciences* 43, no. 1 (2012): 1–3; Sabina Leonelli, *Data-Centric Biology: A Philosophical Study* (Chicago: University of Chicago Press, 2016).

14. Bruno J. Strasser, "Data-Driven Sciences: From Wonder Cabinets to Electronic Databases," *Studies in History and Philosophy of Biological and Biomedical Sciences* 43 (2012): 85–87. For relevant articles in the special issue, see Staffan Müller-Wille and Isabelle Charmantier, "Natural History and Information Overload: The Case of Linnaeus," *Studies in History and Philosophy of Biological and Biomedical Sciences* 43, no. 1 (2012): 4–15; Sabina Leonelli and Rachel A. Ankeny, "Re-Thinking Organisms: The Impact of Databases on Model Organism Biology," *Studies in History and Philosophy of Biological and Biomedical Sciences* 43, no. 1 (2012): 29–36; Peter Keating and Alberto Cambrosio, "Too Many Numbers: Microarrays in Clinical Cancer Research," *Studies in History and Philosophy of Biological and Biomedical Sciences* 43, no. 1 (2012): 37–51.

15. Strasser, "Data-Driven Sciences," 85.

16. Müller-Wille and Charmantier, "Natural History and Information Overload."

17. Strasser, "Data-Driven Sciences," 85.

18. Strasser, "Data-Driven Sciences," 86, 87.

19. Leonelli, *Data-Centric Biology.*

20. To be clear, archeologists, paleontologists, and curators are vital to the pursuit of DNA from ancient and extinct organisms. These researchers' skills are valuable for sample access as well as knowledge of the historical and biological background needed in order to give context to the data obtained from a specific sample. But there is a tension between those researchers responsible for conserving specimen collections and those interested in sampling organisms for genetic information because sampling for ancient DNA is destructive to the specimen. This was certainly a concern in the early years of ancient DNA research. Museums value their collections for their rarity, and their main mission is to conserve past and present specimens for future generations to study or enjoy. While molecular methods offer opportunities for curators to make new uses of old collections, taking samples of skin, tissue, or bone can damage often rare or important specimens. This presents a clear challenge to researchers and curators to find a compromise between their motives. To a certain extent, this challenge can cause a significant dichotomy between the large labs in ancient DNA research that are driving more and more specimen sampling and those curators who are trying to minimize damage to museum collections. For more information, see J. Freedman, Lucy van Dorp, and Selina

Brace, "Destructive Sampling Natural Science Collections: An Overview for Museum Professionals and Researchers," *Journal of Natural Science Collections* 5 (2017): 1–14.

21. This business attitude toward the search for DNA from fossils is not without its critics. Archeologists, paleontologists, and curators are some of the most concerned about the intensity of sampling. See Ewen Callaway, "Divided by DNA: The Uneasy Relationship Between Archaeology and Ancient Genomics," *Nature* 555, no. 7698 (2018): 573–76.

22. Jane Calvert, "Systems Biology, Synthetic Biology and Data-Driven Research: A Commentary on Krohs, Callebaut, and O'Malley and Soyer," *Studies in History and Philosophy of Biological and Biomedical Sciences* 43, no. 1 (2012): 81–84; Ulrich Krohs, "Convenience Experimentation," *Studies in History and Philosophy of Biological and Biomedical Sciences* 43, no. 1 (2012): 52–57; Werner Callebaut, "Scientific Perspectivism: A Philosopher of Science's Response to the Challenge of Big Data Biology," *Studies in History and Philosophy of Biological and Biomedical Sciences* 43, no. 1 (2012): 69–80; Maureen A. O'Malley and Orkun S. Soyer, "The Roles of Integration in Molecular Systems Biology," *Studies in History and Philosophy of Biological and Biomedical Sciences* 43, no. 1 (2012): 58–68.

23. Krohs, "Convenience Experimentation," 53; Richard M. Burian, "Exploratory Experimentation and the Role of Histochemical Techniques in the Work of Jean Brachet, 1938–1952," *History and Philosophy of the Life Sciences* 19, no. 1 (1997): 27–45; Friedrich Steinle, "Entering New Fields: Exploratory Uses of Experimentation," *Philosophy of Science* 64 (1997): S65–74; Leonelli and Ankeny, "Re-Thinking Organisms."

24. Maureen A. O'Malley, "Exploratory Experimentation and Scientific Practice: Metagenomics and the Proteorhodopsin Case," *History and Philosophy of the Life Sciences* 29, no. 3 (2007): 345.

25. "Neandertal Genome to Be Deciphered," Max Planck Society, July 20, 2006, www.mpg.de/534422/pressRelease20060720.

26. Stephen Hilgartner, "Staging High-Visibility Science: Media Orientation in Genome Research," in *The Sciences' Media Connection—Public Communication and Its Repercussions,* ed. Simone Rödder, Martina Franzen, and Peter Weingart (Dordrecht, Netherlands: Springer, 2012), 190, 212.

CHAPTER 10. *JURASSIC PARK* EFFECT

1. Ben Macintyre, "The Great Auk Needn't Be as Dead as a Dodo," *The Times* (London), March 8, 2013, www.thetimes.co.uk/article/the-great-auk-neednt-be -as-dead-as-a-dodo-5gztjkjtrbb.

2. Hendrik Poinar, "Not All Mammoths Were Woolly" (Washington, D.C.: TEDxDeExtinction, 2013), http://reviverestore.org/events/tedxdeextinction/; Beth Shapiro, "Ancient DNA: What It Is and What It Could Be" (Washington, D.C.: TEDxDeExtinction, 2013), http://reviverestore.org/events/tedxdeextinction/.

3. Ben Novak, "How to Bring Passenger Pigeons All the Way Back" (Washington, D.C.: TEDxDeExtinction, 2013), http://reviverestore.org/events/ tedxdeextinction/.

4. John Travis, "Making the Cut," *Science* 350, no. 6267 (2015): 1456–57.

5. Jacob S. Sherkow and Hank T. Greely, "What If Extinction Is Not Forever?" *Science* 5, no. 340 (2013): 32–33.

6. D. T. Max, "Can You Revive an Extinct Animal?" *New York Times*, January 1, 2006, www.nytimes.com/2006/01/01/magazine/01taxidermy.html?pagewanted=all&_r=0.

7. J. Folch et al., "First Birth of an Animal from an Extinct Subspecies (Capra Pyrenaica Pyrenaica) by Cloning," *Theriogenology* 71, no. 6 (2009): 1026–34.

8. "The Great Passenger Pigeon Comeback," Revive & Restore, 2016, http://reviverestore.org/projects/the-great-passenger-pigeon-comeback/.

9. "Woolly Mammoth Revivalists," Revive & Restore, 2016, http://longnow.org/revive/projects/woolly-mammoth/woolly-mammoth-revivalists/.

10. Sasha Harris-Lovett, "'Jurassic World' Paleontologist Wants to Turn a Chicken into a Dinosaur," *Los Angeles Times*, June 12, 2015, www.latimes.com/science/sciencenow/la-sci-sn-horner-dinosaurs-20150612-story.html; Jack Horner and James Gorman, *How to Build a Dinosaur: Extinction Doesn't Have to Be Forever* (New York: Dutton, 2009).

11. Charles Darwin, *On the Origin of Species by Means of Natural Selection; or, The Preservation of Favoured Races in the Struggle for Life* (London: Murray, 1859); Ernst Mayr, *Animal Species and Evolution* (Cambridge, Mass.: Harvard University Press, 1963); David Hull, "The Effect of Essentialism on Taxonomy: Two Thousand Years of Stasis," *British Journal for the Philosophy of Science* 16, no. 16 (1965): 1–18; John Beatty, "Speaking of Species: Darwin's Strategy," in *The Darwinian Heritage*, ed. D. Kohn (Princeton, N.J.: Princeton University Press, 1985); John Dupré, *The Disorder of Things: Metaphysical Foundations of the Disunity of Science* (Cambridge, Mass.: Harvard University Press, 1993); Jody Hey, "The Mind of the Species Problem," *Trends in Ecology and Evolution* 16, no. 7 (2001): 326–29.

12. Alissa Greenberg, "A Brief Look at the Ethical Debate of De-Extinction," *Stanford-Brown International Genetically Engineered Machine Workshop Report* (2013): 1–8, http://2013.igem.org/wiki/images/8/8f/De-Extinction_Ethics.pdf; Sherkow and Greely, "What If Extinction Is Not Forever?" 32.

13. Henry Nicholls, "Let's Make a Mammoth," *Nature* 456 (November 20, 2008): 310–14.

14. Beth Shapiro, *How to Clone a Mammoth: The Science of De-Extinction* (Princeton, N.J.: Princeton University Press, 2015).

15. Mac Margolis, "A Real-Life *Jurassic Park*," *Newsweek*, January 29, 2006, www.newsweek.com/real-life-jurassic-park-108597.

16. Julian Ryall, "DNA Scholars Hope to Stock Siberia 'Park' with Mammoths," *Japan Times* (Tokyo), August 20, 2002, www.japantimes.co.jp/news/2002/08/20/national/dna-scholars-hope-to-stock-siberia-park-with-mammoths/#.WJXl_7GcagQ; Stefan Lovegren, "Woolly Mammoth Resurrection, 'Jurassic Park' Planned," *National Geographic News*, April 8, 2005, www.nationalgeographic.com/pages/topic/latest-stories; Sergey A. Zimov,

"Pleistocene Park: Return of the Mammoth's Ecosystem," *Science* 308, no. 5723 (2005): 796–98.

17. Margolis, "A Real-Life *Jurassic Park*."

18. Michael Archer, "Second Chance for Tasmanian Tigers and Fantastic Frogs" (Washington, D.C.: TEDxDeExtinction, 2013), http://reviverestore.org/events/tedxdeextinction/.

19. Don Colgan and Mike Archer, "The Thylacine Project," *Australasian Science* 21, no. 1 (2000): 21.

20. Amy Lynn Fletcher, "Bring 'Em Back Alive: Taming the Tasmanian Tiger Cloning Project," *Technology in Society* 30, no. 2 (2008): 194–201; Amy Fletcher, "Genuine Fakes: Cloning Extinct Species as Science and Spectacle," *Politics and the Life Sciences* 29, no. 1 (2010): 48–60.

21. Fletcher, "Genuine Fakes," 51–52.

22. Margolis, "A Real-Life *Jurassic Park*."

23. Craig D. Millar et al., "New Developments in Ancient Genomics," *Trends in Ecology and Evolution* 23, no. 7 (2008): 386–93.

24. Nicolas Wade, "Regenerating a Mammoth for $10 Million," *New York Times*, International Edition, November 19, 2008, www.nytimes.com/2008/11/20/science/20mammoth.html?_r=0.

25. Philip Bethge and Johann Grolle, "Can Neanderthals Be Brought Back from the Dead?" *Der Spiegel*, January 18, 2013, www.spiegel.de/international/zeitgeist/george-church-explains-how-dna-will-be-construction-material-of-the-future-a-877634.html.

26. Susan Young Rojahn, "Wanted: Surrogate for Neanderthal Baby," *MIT Technology Review*, January 17, 2013, www.technologyreview.com/s/510071/wanted-surrogate-for-neanderthal-baby/.

27. Fiona Macrae, "'Adventurous Human Woman' Wanted to Give Birth to Neanderthal Man by Harvard Professor," *Daily Mail* (London), January 20, 2013, www.dailymail.co.uk/news/article-2265402/Adventurous-human-woman-wanted-birth-Neanderthal-man-Harvard-professor.html.

28. "Neanderthal Baby Clone: George Church, Harvard Geneticist, Looks to Resurrect Extinct Species," Huffington Post Canada, January 21, 2013, www.huffingtonpost.ca/2013/01/21/neanderthal-baby-clone_n_2521027.html.

29. Svante Pääbo, "Neanderthals Are People, Too," *New York Times*, April 24, 2014, www.nytimes.com/2014/04/25/opinion/neanderthals-are-people-too.html.

30. "Surrogate Mother (Not Yet) Sought for Neanderthal," *Der Spiegel*, January 23, 2013, www.spiegel.de/international/spiegel-responds-to-brouhaha-over-neanderthal-clone-interview-a-879311.html.

31. Dorothy Nelkin, *Selling Science: How the Press Covers Science and Technology* (New York: W. H. Freeman, 1995), 162.

32. Quoted from Nelkin, *Selling Science*, 82.

33. Sharon Dunwoody, "Science Journalism: Prospects of the Digital Age," in *Routledge Handbook of Public Communication of Science and Technology*, 2nd edition, ed. Massimiano Bucchi and Brian Trench (London: Routledge, 2014), 32.

34. Martina Franzen, *Breaking News: Wissenschaftliche Zeitschriften Im Kampf Um Aufmerksamkeit* (Baden-Baden, Germany: Nomos, 2011).

35. Sally Gregory Kohlstedt, *The Formation of the American Scientific Community: The American Association for the Advancement of Science, 1848–1860* (Champaign: University of Illinois Press, 1976); Sally Gregory Kohlstedt, Michael Sokal, and Bruce V. Lewenstein, *The Establishment of Science in America: 150 Years of the American Association for the Advancement of Science* (New Brunswick, N.J.: Rutgers University Press, 1999); Melinda Baldwin, *Making "Nature": The History of a Scientific Journal* (Chicago: University of Chicago Press, 2015).

36. Peter Weingart, "The Lure of the Mass Media and Its Repercussions on Science," in *The Sciences' Media Connection—Public Communication and Its Repercussions,* ed. Simone Rödder, Martina Franzen, and Peter Weingart (Dordrecht, Netherlands: Springer, 2012), 29.

37. Michael B. Shermer, "This View of Science: Stephen Jay Gould as Historian of Science and Scientific Historian, Popular Scientist and Scientific Popularizer," *Social Studies of Science* 32, no. 4 (2002): 490. See also Keay Davidson, *Carl Sagan: A Life* (New York: Wiley, 1999).

38. Graeme Turner, *Understanding Celebrity* (London: SAGE, 2004), 3.

39. Neil Hall, "The Kardashian Index: A Measure of Discrepant Social Media Profile for Scientists," *Genome Biology* 15, no. 424 (2014): 1–2.

40. Nelkin, *Selling Science*, 13.

41. Massimiano Bucchi, "When Scientists Turn to the Public: Alternative Routes in Science Communication," *Public Understanding of Science* 5, no. 4 (1996): 375–94.

42. Bruce V. Lewenstein, "From Fax to Facts: Communication in the Cold Fusion Saga," *Social Studies of Science* 25 (1995): 403–36; Bart Simon, *Undead Science: Science Studies and the Afterlife of Cold Fusion* (New Brunswick, N.J.: Rutgers University Press, 2002).

43. Angela Cassidy, "Popular Evolutionary Psychology in the UK: An Unusual Case of Science in the Media?" *Public Understanding of Science* 14 (2005): 115–41; Angela Cassidy, "Evolutionary Psychology as Public Science and Boundary Work," *Public Understanding of Science* 15 (2006): 175–205.

44. Felicity Mellor, "Colliding Worlds: Asteroid Research and the Legitimization of War in Space," *Social Studies of Science* 37, no. 4 (2007): 499–531; Felicity Mellor, "Negotiating Uncertainty: Asteroids, Risk and the Media," *Public Understanding of Science* 19, no. 1 (2010): 16–33.

45. Rae Goodell, *The Visible Scientists* (Boston: Little, Brown, 1977); Jane Gregory, *Fred Hoyle's Universe* (Oxford: Oxford University Press, 2005); Declan Fahy, *The New Celebrity Scientists: Out of the Lab and into the Limelight* (Lanham, Md.: Rowman and Littlefield, 2015).

46. Fletcher, "Genuine Fakes," 49.

47. Rödder, Franzen, and Weingart, eds., *The Sciences' Media Connection.*

48. Zimov, "Pleistocene Park."

49. Sharon M. Friedman, Sharon Dunwoody, and Carol L. Rogers, eds., *Scientists and Journalists: Reporting Science as News* (New York: Free Press, 1986), xiii.

50. Sharon Dunwoody attributed this idea to Blumler and Gurevitch. See Sharon Dunwoody, "The Scientist as Source," in Friedman, Dunwoody, and Rogers, eds., *Scientists and Journalists*, 13. See also Jay G. Blumler and Michael Gurevitch, "Politicians and the Press: An Essay on Role Relationships," in *Handbook of Political Communication*, ed. Dan D. Nimmo and Keith R. Sanders (Beverly Hills, Calif.: Sage, 1981), 467–93.

51. Peter Broks, *Understanding Popular Science* (Maidenhead, U.K.: Open University Press, 2006), 144. The use of the term "popular science" has a contentious history. Recently, James Secord has argued for abandoning it. See James A. Secord, "Knowledge in Transit," *Isis* 95, no. 4 (2004): 654–72. Broks has reconsidered the term by redefining how we think and talk about popular science.

52. Broks, *Understanding Popular Science*, 107, 149.

53. Nelkin, *Selling Science*, 145.

54. "Media Frenzy," *Nature* 459, no. 7246 (2009): 484.

EPILOGUE: ANCIENT DNA AS CELEBRITY SCIENCE

1. Alida M. Bailleul et al., "Evidence of Proteins, Chromosomes and Chemical Markers of DNA in Exceptionally Preserved Dinosaur Cartilage," *National Science Review* 7, no. 4 (2020): 815–22.

2. Josh K. Elliott, "Bingo? Possible Dinosaur 'DNA' Found in 75-Million-Year-Old Fossil," *Global News* (Canada), March 4, 2020, https://globalnews.ca/news/6625164/dinosaur-dna-found-fossil/#:~:text=The%20Jurassic%20Park%20dream%20has,belonged%20to%20a%20baby%20dinosaur.

3. Michael Greshko, "Hints of Fossil DNA Discovered in Dinosaur Skull," *National Geographic*, March 3, 2020, www.nationalgeographic.com/science/2020/03/hints-of-dna-discovered-in-a-dinosaur-fossil/.

4. Ludovic Orlando et al., "Recalibrating Equus Evolution Using the Genome Sequence of an Early Middle Pleistocene Horse," *Nature* 499, no. 7456 (2013): 74–78.

5. Joe Hansen, "700,000-Year-Old Horse Genome Shatters Record for Sequencing of Ancient DNA," *Wired*, 2013, www.wired.com/2013/06/ancient-horse-genome/.

6. Robin McKie, "Prehistoric DNA Sequencing: *Jurassic Park* Was Not So Wide of the Mark," *The Guardian* (London), July 6, 2013, www.theguardian.com/science/2013/jul/07/prehistoric-horse-dna-genome-sequence.

7. Kate Wong, "Mammoth Genomes Shatter Record for Oldest DNA Sequences," *Scientific American*, February 17, 2021, www.scientificamerican.com/article/mammoth-genomes-shatter-record-for-oldest-dna-sequences/.

8. Tom van der Valk et al., "Million-Year-Old DNA Sheds Light on the Genomic History of Mammoths," *Nature* 591, no. 7849 (2021): 265–69.

9. Belinda Smith, "Million-Year-Old DNA from Mammoth Teeth Found in Siberia Is Oldest Genome Ever Sequenced," ABC Science, February 17, 2021, www.abc.net.au/news/science/2021-02-18/mammoth-woolly-dna-siberia-russia-palaeogenetics-permafrost/13160930.

10. Rae Goodell, *The Visible Scientists* (Boston: Little, Brown, 1977); Jane Gregory and Steve Miller, *Science in Public: Communication, Culture, and Credibility* (Cambridge, Mass.: Basic, 1998); Peter Broks, *Understanding Popular Science* (Maidenhead, U.K.: Open University Press, 2006); Declan Fahy, *The New Celebrity Scientists: Out of the Lab and into the Limelight* (Lanham, Md.: Rowman and Littlefield, 2015).

11. William S. Pretzer, *Working at Inventing: Thomas A. Edison and the Menlo Park Experience* (Dearborn, Mich.: Henry Ford Museum and Greenfield Village, 1989); John D. Barrow, "Einstein as Icon," *Nature* 433, no. 7023 (2005): 218–19.

12. Keay Davidson, *Carl Sagan: A Life* (New York: Wiley, 1999); Jane Gregory, *Fred Hoyle's Universe* (Oxford: Oxford University Press, 2005).

13. Patricia Fara, *Newton: The Making of a Genius* (New York: Columbia University Press, 2003).

14. Sharon M. Friedman, Sharon Dunwoody, and Carol L. Rogers, eds., *Scientists and Journalists: Reporting Science as News* (New York: Free Press, 1986); Gregory and Miller, *Science in Public*; Broks, *Understanding Popular Science*.

15. Goodell, *The Visible Scientists*, 264. Goodell argued that visible scientists embody personal and professional traits that make them attractive to the public. She suggested that visible scientists are articulate, have a colorful image as well as a credible reputation, and speak on hot or controversial topics.

16. Sean Connor, "Boy from Bingley 'Lobbed Intellectual Grenades' at Science," *The Independent* (London), August 23, 2001, 11; N. Martel, "Mysteries of Life, Time and Space (and Green Slime)," *New York Times*, September 28, 2004, E5; "Popularizer Greenfield Is Blackballed by Peers," *Nature* 429 (2004): 9; J. Bohannon, "The Baroness and the Brain," *Science* 310, no. 5750 (2005): 962; Declan Fahy and Bruce Lewenstein, "Scientists in Popular Culture: Making Celebrities," in *Routledge Handbook of Public Communication of Science and Technology*, 2nd edition, ed. Massimiano Bucchi and Brian Trench (London: Routledge, 2014), 87.

17. Fahy and Lewenstein, "Scientists in Popular Culture," 86, 93.

18. Fahy, *The New Celebrity Scientists*, 3.

19. Chris Rojek, *Celebrity* (London: Reaktion, 2001); Jessica Evans and David Hesmondhalgh, eds., *Understanding Media: Inside Celebrity* (Maidenhead, U.K.: Open University Press, 2005); Graeme Turner, *Understanding Celebrity* (London: Sage, 2004).

20. Turner, *Understanding Celebrity*, 9.

21. Quoted in Turner, *Understanding Celebrity*, 5.

22. Rojek, *Celebrity*, 10, 17–18.

23. I focus on the Public Understanding of Science (PUS) movement in the United Kingdom because it was the initial impetus behind the popularization of science and technology at this time. However, this is not a complete treatment of this movement or other movements elsewhere. See Gregory and Miller, *Science in Public*, for more information. In order to fully understand the context in which a celebrity science like ancient DNA research could have evolved, and indeed did, analyses of international science communication movements are necessary. On this note, it is interesting to mention ancient DNA researchers from different countries (United Kingdom, United States, and Canada to Germany, Denmark,

and Australia) face different political pressures, as well as science and science communication traditions. Regardless of these differences, however, they all felt, and continue to feel, the pressure to publicize their science to the broader public. This is likely because many ancient DNA researchers are seeking high-profile publications in journals like *Nature* and *Science* that come from Anglo-Saxon traditions, which are in turn influenced by the U.K. and U.S. science communication movements. The extent of the effects of these movements and others requires further research.

24. Walter Bodmer, "The Public Understanding of Science," report published by the Royal Society, London, 1985, 1–41; Gregory and Miller, *Science in Public*, 1–18; Broks, *Understanding Popular Science*, 96–117.

25. Gregory and Miller, *Science in Public*, 19–45.

26. Broks, *Understanding Popular Science*, 107.

27. Gregory and Miller, *Science in Public*, 2.

28. John C. Burnham, *How Superstition Won and Science Lost: Popularizing Science and Health in the United States* (New Brunswick, N.J.: Rutgers University Press, 1987); Jan Golinski, *Science as Public Culture: Chemistry and Enlightenment in Britain, 1760–1820* (Cambridge: Cambridge University Press, 1992); Evans and Hesmondhalgh, eds., *Understanding Media*; Simone Rödder, Martina Franzen, and Peter Weingart, eds., *The Sciences' Media Connection—Public Communication and Its Repercussions* (Dordrecht, Netherlands: Springer, 2012); Sharon Dunwoody, "Science Journalism: Prospects of the Digital Age," in Bucchi and Trench, eds., *Routledge Handbook of Public Communication of Science and Technology*, 27–39.

29. Rödder, Franzen, and Weingart, eds., *The Sciences' Media Connection*.

30. Evans and Hesmondhalgh, eds., *Understanding Media*, 12.

31. Broks, *Understanding Popular Science*, 107, 149.

32. Elisabeth S. Clemmens, "Of Asteroids and Dinosaurs: The Role of the Press in the Shaping of Scientific Debate," *Social Studies of Science* 16 (1986): 421–56; Elisabeth S. Clemmens, "The Impact Hypothesis and Popular Science: Conditions and Consequences of Interdisciplinary Debate," in *The Mass-Extinction Debates: How Science Works in a Crisis*, ed. William Glen (Stanford, Calif.: Stanford University Press, 1994), 92–120; Bruce V. Lewenstein, "From Fax to Facts: Communication in the Cold Fusion Saga," *Social Studies of Science* 25 (1995): 403–36; Angela Cassidy, "Popular Evolutionary Psychology in the UK: An Unusual Case of Science in the Media?" *Public Understanding of Science* 14 (2005): 115–41; Angela Cassidy, "Evolutionary Psychology as Public Science and Boundary Work," *Public Understanding of Science* 15 (2006): 175–205; Felicity Mellor, "Colliding Worlds: Asteroid Research and the Legitimization of War in Space," *Social Studies of Science* 37, no. 4 (2007): 499–531; Felicity Mellor, "Negotiating Uncertainty: Asteroids, Risk and the Media," *Public Understanding of Science* 19, no. 1 (2010): 16–33; Stephen Hilgartner, "Staging High-Visibility Science: Media Orientation in Genome Research," in *The Sciences' Media Connection—Public Communication and Its Repercussions*, ed. Simone Rödder, Martina Franzen, and Peter Weingart (Dordrecht, Netherlands: Springer, 2012), 189–215; Rödder, Franzen, and Weingart, eds., *The Sciences' Media Connection*.